Structure Topology and Symplectic Geometry

Edited by Kamal Shah and Min Lei

Published in London, United Kingdom

IntechOpen

Supporting open minds since 2005

Structure Topology and Symplectic Geometry
http://dx.doi.org/10.5772/intechopen.80177
Edited by Kamal Shah and Min Lei

Contributors
Cenap Özel, Mesfer Al Qahtani, Yoritaka Iwata, Ying-Qiu Gu, Min Lei, Samuel Krushkal, Karthic Mohan, Jananeeswari Narayanamoorthy, Ibtesam Alshammari

First published in London, United Kingdom, 2021 by IntechOpen
IntechOpen is the global imprint of INTECHOPEN LIMITED, registered in England and Wales, registration number: 11086078, 5 Princes Gate Court, London, SW7 2QJ, United Kingdom
Printed in Croatia

British Library Cataloguing-in-Publication Data
A catalogue record for this book is available from the British Library

Additional hard and PDF copies can be obtained from orders@intechopen.com

Structure Topology and Symplectic Geometry
Edited by Kamal Shah and Min Lei
p. cm.
Print ISBN 978-1-83962-598-5
Online ISBN 978-1-83962-599-2
eBook (PDF) ISBN 978-1-83962-600-5

For EU product safety concerns:
IN TECH d.o.o., Prolaz Marije Krucifikse Kozulić 3, 51000 Rijeka, Croatia, info@intechopen.com or visit our website at intechopen.com.

Meet the editors

Dr. Kamal Shah is an associate professor in the Department of Mathematics, University of Malakand, Chakdara Dir (L), Khyber Pakhtan Khawa, Pakistan. He obtained an MSc from Government Post Graduate Jehanzeb College Swat in 2005. He joined the University of Malakand as a lecturer in 2007. Dr. Shah obtained a Ph.D. in Fractional Calculus in 2016 from the University of Malakand. He has published numerous articles in scientific journals and has supervised many MPhil and Ph.D. students.

Dr. Min Lei has worked at Shanghai Jiaotong University as an associate professor since 2004. From February 2003 to February 2004, she worked as a research fellow in the Department of Mechanical Engineering, Nanyang University of Technology, Singapore. From 2000 to 2002, she worked as a postdoctoral fellow in the Institute of Mechanical and Electrical Control, School of Mechanical and Power Engineering, Shanghai Jiaotong University. She received a doctoral degree from the Department of Biomedical Engineering, Shanghai Jiaotong University, in 2000. Dr. Lei has long been engaged in biomedical signal processing, modern signal processing, pattern recognition and classification, and nonlinear dynamics.

Contents

Preface **XI**

Chapter 1 **1**
A Review Note on the Applications of Linear Operators in Hilbert Space
by Karthic Mohan and Jananeeswari Narayanamoorthy

Chapter 2 **19**
Operator Topology for Logarithmic Infinitesimal Generators
by Yoritaka Iwata

Chapter 3 **27**
New Topology on Symmetrized Omega Algebra
by Mesfer Alqahtani, Cenap Özel and Ibtesam Alshammari

Chapter 4 **43**
Some Applications of Clifford Algebra in Geometry
by Ying-Qiu Gu

Chapter 5 **69**
Quasiconformal Reflections across Polygonal Lines
by Samuel L. Krushkal

Chapter 6 **103**
Symplectic Geometry and Its Applications on Time Series Analysis
by Min Lei

Preface

Topology is the science of position and surfaces. It endows with a structure. Symplectic geometry is an even-dimensional geometry living on an even-dimensional structure space. Generally, topology has many applications in a variety of disciplines including networking, communications, dynamics, and other fields of science. Structure topology is an interesting branch of general topology. Symplectic geometry studies the invariant properties of structure graphics after symplectic transformation. It has beautiful mathematical principles and contents such as the duality principle, continuity principle, and butterfly theorem. With the development of research in the field, symplectic geometry is used not only to solve classical nonlinear dynamical systems but also to address various nonlinear time series on physics, engineering, and biomedical engineering.

Chapter 1 is a review of the applications of linear operators and Chapter 2 presents interesting results about operators working on topologies. Chapter 3 establishes some detail about the use of topology on symmetrized Omega algebra. Chapter 4 discusses applications of Clifford algebra in geometry, while Chapter 5 studies quasiconformal reflection across polygonal lines. Finally, Chapter 6 introduces some basic concepts and elements of mathematics in symplectic geometry theory. Algorithms based on MATLAB software are provided for applications of symplectic geometry on time series analysis, such as embedding dimension estimation, nonlinear testing, noise reduction, and fault diagnosis.

This book will help researchers and students understand how topology and structures are related and how to apply symplectic geometry theory to analyze practical data from engineering fields. The content of this book is conducive to the development and applications of structure topology and symplectic geometry in other branches of sciences and engineering.

Kamal Shah
Associate Professor,
Department of Mathematics,
University of Malakand,
Chakdara Dir(Lower), Khyber Pakhtankhawa, Pakistan

Min Lei
Associate Professor,
School of Mechanical Engineering,
Shanghai Jiao Tong University,
Shanghai, China

Chapter 1

A Review Note on the Applications of Linear Operators in Hilbert Space

Karthic Mohan and Jananeeswari Narayanamoorthy

Abstract

Hilbert Spaces are the closest generalization to infinite dimensional spaces of the Euclidean Spaces. We Consider Linear transformations defined in a normed space and we see that all of them are Continuous if the Space is finite Dimensional Hilbert Space Provide a user-friendly framework for the study of a wide range of subjects from Fourier Analysis to Quantum Mechanics. The adjoint of an Operator is defined and the basic properties of the adjoint operation are established. This allows the introduction of self Adjoint Operators are the subject of the section.

Keywords: linear space, norm of a vector, inner product, orthogonal vector, adjoint of an operator

1. Introduction

The Concept of Hilbert Space was put forwarded by David Hilbert in his work on Quadratic forms in infinitely many Variables. We take a Closer look at Linear Continuous map between Hilbert Spaces [1]. These are called bounded operators and branch of Functional Analysis Called "Operator Theory" [2]. Next we derive an important inequality which has many interesting applications in the theory of inner product spaces and as a consequence we obtain that each inner product space is a normed Vector spaces with the norm [3], i.e. the inner product generates this form. Moreover there are several essential algebraic identities, variously and ambiguously called Polarization Identities. These and other closely related identities are of constant use. Now we are in position to state and prove the above mentioned important inequality known as Cauchy-Schwartz Buniakowsi inequality (briefly we say CSB inequality) and we shall also use this to define the concept of angle by means of a formula [1]. The theory of Hilbert Space that Hilbert and Others developed has not only greatly enriched the world of Mathematics but has Proven Extremely useful in the development of Scientific Theories Particularly Quantum Mechanics.

2. Definition 1

A Hilbert space is a complex Banach space whose norm induced from an inner produce [4] i.e., in which there is defined a complex function (x,y) of vectors x & y and α, β are scalars with the following properties

i. $(\alpha x + \beta y, z) = \alpha(x, z) + \beta(y, z)$

ii. $(\overline{x, y}) = (y, x)$

iii. $(x, x) = \|x\|^2$

2.1 Remark 1.1

Every polynomial equation of the nth degree with complex co-efficient has exactly n complex roots [5].

In accordance with the above remarks the scalars in this example are understood to be the complex number.

Consider the space l_2^n with the inner product of two vectors $x = (x_1, x_2, \dots, x_n)$ and $y = (y_1, y_2, \dots, y_n)$ defined by

$$(x, y) = \sum_{i=1}^{n} x_y \overline{y}_i$$

Now we are going to prove that l_2^n is a Hilbert space.

2.1.1 Proof

By using Hilbert Space definition $[(x, y)$ of complex function, α, β are scalars$]$

i. $(\alpha x + \beta y, z) = \sum_{i=1}^{n} (\alpha x_i + \beta y_i) \overline{z}_i$

$= \alpha \sum_{i=1}^{n} x_i \overline{z}_i + \beta \sum_{i=1}^{n} y_i \overline{z}_i$

$(\alpha x + \beta y, z) = \alpha(x, z) + \beta(y, z) \forall x, y, z \in l_2^n$

ii. $(\overline{x, y}) = \sum_{i=1}^{n} \overline{x_i \overline{y}_i}$

$= \sum_{i=1}^{n} \overline{x}_i \overline{\overline{y}}_i$

$= \sum_{i=1}^{n} y_i \overline{x}_i$

$= (y, x)$

iii. $(x, x) = \sum_{i=1}^{n} x_i \overline{x}_i$

$= \sum_{i=1}^{n} |x_i|^2$

$(x, x) = \|x\|^2$

Therefore (x, y) is an inner product on l_2^n.

Therefore l_2^n is a Hilbert space.

2.2 Theorem 1.1 (Schwartz inequality)

If x and y are any two vectors in Hilbert space then $|(x, y)| \leq \|x\| \|y\|$ [6–8].

2.2.1 Proof

When $y = 0$, the result is clear for both sides vanish

DOI: http://dx.doi.org/10.5772/intechopen.92773

i.e., $|(x, y)| = |(x, 0)| = |0| = 0$

$$\|x\|\|y\| = \|x\|\|0\| = \|x\|0 = 0$$
$$|(x, y)| = \|x\|\|y\| = 0$$

When $y \neq 0$
Take any scalar $\alpha \in C$ [Complex banach space] always $\|x - \alpha y\|^2 \geq 0$

$$(x - \alpha y, x - \alpha y) \geq 0$$
$$(x, x) - (x, \alpha y) - (\alpha y, x) + (\alpha y, \alpha y) \geq 0$$
$$(x, x) - \overline{\alpha}(x, y) - \alpha(y, x) + \alpha\overline{\alpha}(y, y) \geq 0$$
$$\|x\|^2 - \overline{\alpha}(x, y) - \overline{\alpha(x, y)} + \alpha\overline{\alpha}\|y\|^2 \geq 0 \tag{1}$$

Put $\alpha = \frac{(x,y)}{\|y\|^2}$ where $(x, y) \in C$.
$\because y \neq 0$ and $\|y\| \neq 0$.
So choose $\alpha = \frac{(x,y)}{\|y\|^2}, \overline{\alpha} = \frac{\overline{(x,y)}}{\|y\|^2}$.
From Eq. (1) becomes

$$\|x\|^2 - \frac{\overline{(x, y)}}{\|y\|^2}(x, y) - \frac{(x, y)}{\|y\|^2}\overline{(x, y)} + \frac{(x, y)}{\|y\|^2}\frac{\overline{(x, y)}}{\|y\|^2}\|y\|^2 \geq 0$$
$$\|x\|^2 - \frac{|(x, y)|^2}{\|y\|^2} - \frac{|(x, y)|^2}{\|y\|^2} + \frac{|(x, y)|^2}{\|y\|^2} \geq 0$$
$$\|x\|^2 - \frac{|(x, y)|^2}{\|y\|^2} \geq 0$$
$$\frac{|(x, y)|^2}{\|y\|^2} \leq \|x\|^2$$
$$|(x, y)|^2 \leq \|x\|^2\|y\|^2$$
$$\therefore |(x, y)| \leq \|x\|\|y\|$$

2.3 Result 1

An inner product space is a normal linear space [9].

2.3.1 Proof

To prove.
$\|x\| \geq 0$ and $\|x\| = 0$ if $x = 0$

$$\|x\| = \sqrt{(x, x)} \Rightarrow \|x\|^2 = (x, x) \geq 0$$

So that $\|x\| \geq 0$ and $\|x\| = 0 \Leftrightarrow x = 0$.
Now we have to show that $\|x + y\| \leq \|x\| + \|y\|$

$$\|x + y\|^2 = [(x + y), (x + y)]$$
$$= (x, x) + (x, y) + (y, x) + (y, y)$$

$$= \|x\|^2 + (x,y) + \overline{(x,y)} + \|y\|^2$$
$$= \|x\|^2 + 2\,Re\,(x,y) + \|y\|^2$$
$$\leq \|x\|^2 + 2\|x\|\|y\| + \|y\|^2$$
$$\|x+y\|^2 \leq (\|x\| + \|y\|)^2$$
$$\|x+y\| \leq \|x\| + \|y\|$$

Now we can prove that $\|\alpha x\| = |\alpha|\|x\|$.
Consider

$$\|\alpha x\|^2 = (\alpha x, \alpha x)$$
$$= \alpha\bar{\alpha}(x,x)$$
$$\|\alpha x\|^2 = |\alpha|^2\|x\|^2$$
$$\|\alpha x\| = |\alpha|\|x\|$$

An inner product is a normed linear space.

2.4 Result 1.1

An inner product in Hilbert space is jointly continuous [10].

2.4.1 Proof

Since $x_n \to x$ and $y_n \to y \Rightarrow (x_n, y_n) \to (x,y)$
We have.
$\|x_n - x\| \to 0$ as $n \to \infty$ and
$\|y_n - y\| \to 0$ as $n \to \infty$ and
Now consider

$$|(x_n,y_n),(x,y)| = |(x_n,y_n) - (x_n,y) + (x_n,y) - (x,y)|$$
$$\leq |(x_n,y_n) - (x_n,y)| + |(x_n,y) - (x,y)|$$
$$\leq \|x_n\|\|y_n - y\| + \|x_n - x\|\|y\| \text{ (by Schwartz Inequality)}$$
$$\to 0 \text{ as } y_n \to y \text{ and } x_n \to x$$

$|(x_n,y_n),(x,y)| \to 0$ as $n \to \infty$
$(x_n,y_n) \to (x,y)$ as $n \to \infty$

2.5 Theorem 1.1 (parallelogram law in Hilbert space)

If x and y are any two vectors in Hilbert space then

$$\|x+y\|^2 + \|x-y\|^2 = 2\|x\|^2 + 2\|y\|^2 \tag{2}$$

2.5.1 Proof

$$\|x+y\|^2 = (x+y, x+y)$$
$$= (x,x) + (x,y) + (y,x) + (y,y)$$

$$= \|x\|^2 + (x,y) + \overline{(x,y)} + \|y\|^2 \quad (3)$$

$$\|x-y\|^2 = (x-y, x-y)$$
$$= (x,x) - (x,y) - (y,x) - (y,y)$$
$$= \|x\|^2 - (x,y) - \overline{(x,y)} + \|y\|^2 \quad (4)$$

Adding (3) and (4) we get,

$$\|x+y\|^2 + \|x-y\|^2 = 2\|x\|^2 + 2\|y\|^2$$

2.6 Theorem 1.2 (polarization identity)

If x and y are any two vectors in Hilbert space then

$$4(x,y) = \|x+y\|^2 - \|x-y\|^2 + i\|x+iy\|^2 - i\|x+iy\|^2 \quad (5)$$

2.6.1 Proof

$$\|x+y\|^2 = (x+y, x+y)$$
$$= (x,x) + (x,y) + (y,x) + (y,y)$$
$$= \|x\|^2 + (x,y) + \overline{(x,y)} + \|y\|^2 \quad (6)$$
$$\|x-y\|^2 = (x-y, x-y)$$
$$= (x,x) - (x,y) - (y,x) + (y,y)$$
$$= \|x\|^2 - (x,y) - \overline{(x,y)} + \|y\|^2 \quad (7)$$

Subtracting (6) and (7) we get

$$\|x+y\|^2 - \|x-y\|^2 = 2(x,y) + 2(y,x) \quad (8)$$

Replace y by iy in Eq. (8)

$$\|x+y\|^2 - \|x-y\|^2 = 2(x,iy) + 2(iy,x)$$
$$= 2\bar{i}(x,y) + 2i(y,x)$$
$$= -2i(x,y) + 2i(y,x)$$

Multiply both sides by i

$$i\|x+y\|^2 - i\|x-y\|^2 = 2(x,y) - 2(y,x) \quad (9)$$

Adding (8) and (9) we get

$$\|x+y\|^2 - \|x-y\|^2 + i\|x+y\|^2 - i\|x-y\|^2 = 4(x,y). \quad (10)$$

3. Definition 2

Let B be an arbitrary banach space. A convex set in B is a non-empty subset S with the property that if x and y are in S then

$$z = x + t(y - x) = (1 - t)x + ty$$

it also in S for all real number t such that $0 \leq t \leq 1$.

A convex set is a non-empty set which contains the segment joining any pairs of its points.

Since C is convex it is non-empty and contains $\frac{x+y}{2}$ whenever it contains x and y [11].

3.1 Theorem 2.1

3.1.1 Application of parallelogram law

A closed convex subset C of a Hilbert space H contains a unit vector of smallest norm [12].

3.1.2 Proof

Step 1:

Since C is a convex set, it is non empty and contains $\frac{x+y}{2} \in C$ whenever $x, y \in C$.
Let $d = \inf\{\|x\|/x \in C\}$ then

$$d \leq \|x\| \forall x \in C \tag{11}$$

Then there exist a sequence $\{x_n\}$ of vectors in C such that $\{x_n\} \to d$ as

$$n \to \infty \tag{12}$$

Let $x_m, x_n \in C$.
$\because C$ is convex, $\frac{x_m + x_n}{2} \in C$
By (11), $d \leq \left\|\frac{x_m + x_n}{2}\right\|$

$$d \leq \frac{\|x_m + x_n\|}{2}$$

$$2d \leq \|x_m + x_n\|$$

$$4d^2 \leq \|x_m + x_n\|^2$$

$$-4d^2 \geq -\|x_m + x_n\|^2 \tag{13}$$

By parallelogram law

$$\|x_m + x_n\|^2 + \|x_m - x_n\|^2 = 2\|x_m\|^2 + 2\|x_n\|^2$$

$$\|x_m - x_n\|^2 = 2\|x_m\|^2 + 2\|x_n\|^2 - \|x_m + x_n\|^2$$

$$\|x_m - x_n\|^2 \leq 2\|x_m\|^2 + 2\|x_n\|^2 - 4d^2 \text{ by Eq. (13)}$$
$$\leq 2d^2 + 2d^2 - 4d^2 \text{ by Eq. (12) as } n \to \infty.$$
$$\|x_m - x_n\| \to 0 \text{ as } n \to \infty, m \to \infty.$$

$\therefore \{x_n\}$ is a Cauchy sequence in C.
$\because C$ is a closed set in complete Space H
$\Rightarrow C$ is complete
There exist a vector x in C such that $x_n \to x$
i.e.) $x = \lim x_n$

$$\begin{aligned} \|x\| &= \| lim\, x_n \| \\ &= \lim \|x_n\| = d \\ &= \inf \left\{ \frac{\|x\|}{x} \in C \right\} \end{aligned}$$

$\Rightarrow \|x\|$ is smallest.
$\therefore x$ is a vector in C with smallest norm.
Step 2:
To prove uniqueness of x.
Suppose there exist a vector x in C with $\|x'\| = d$ and $x \neq x'$ in C
$\because C$ is convex, $\frac{x+x'}{2} \in C$
By Eq. (11)

$$d \leq \left\| \frac{x + x'}{2} \right\| \tag{14}$$

By parallelogram law,

$$\left\| \frac{x + x'}{2} \right\|^2 + \left\| \frac{x - x'}{2} \right\|^2 = 2\left\| \frac{x}{2} \right\|^2 + 2\left\| \frac{x'}{2} \right\|^2$$

$$\begin{aligned} \left\| \frac{x + x'}{2} \right\|^2 &= 2\left\| \frac{x}{2} \right\|^2 + 2\left\| \frac{x'}{2} \right\|^2 - \left\| \frac{x + x'}{2} \right\|^2 \\ &\leq 2\left\| \frac{x}{2} \right\|^2 + 2\left\| \frac{x'}{2} \right\|^2 \\ &\leq \frac{2\|x\|^2}{2} + \frac{2\|x'\|^2}{2} \\ &\leq \frac{2d^2}{4} + \frac{2d^2}{4} \\ &\leq \frac{4d^2}{4} \\ \left\| \frac{x + x'}{2} \right\|^2 &\leq d^2 \\ \left\| \frac{x + x'}{2} \right\| &\leq d \end{aligned}$$

Which is a contradiction to Eq. (14)
Therefore our assumption on $x \neq x'$ is wrong
Hence $x = x'$

3.2 Theorem 2.2 (orthogonal complements)

Two Vectors x and y in a Hilbert space H are said to be orthogonal (written as $x \perp y$) if $(x, y) = 0$ [9]

i. $x \perp y \Leftrightarrow (\mathrm{x}, \mathrm{y}) = 0$

$\Leftrightarrow (\overline{\mathrm{x}, \mathrm{y}}) = \overline{0}$

$$\Leftrightarrow (y, x) = 0$$

$$\Leftrightarrow y \perp x$$

ii. $(0, x) = 0 \forall x \epsilon H$

$$\Rightarrow 0 \perp x \forall x \epsilon H$$

Therefore 0 is orthogonal to every vector x in H

iii. $x \perp x \Leftrightarrow (\mathrm{x}, \mathrm{x}) = 0$

$$\Leftrightarrow \|x\|^2 = 0$$

$$\Leftrightarrow \|x\| = 0$$

$$\Leftrightarrow x = 0$$

This means that 0 is the only vector orthogonal to itself

3.3 Theorem 2.3 (Pythagorean theorem)

Geometric fact about orthogonal vectors in the Pythagorean theorem such that $x \perp y$ implies [9]

$$\|x + y\|^2 = \|x - y\|^2 = \|x\|^2 + \|y\|^2$$

3.3.1 Proof

Since

$$x \perp y \Leftrightarrow (\mathrm{x}, \mathrm{y}) = 0$$

$$\Leftrightarrow (\overline{\mathrm{x}, \mathrm{y}}) = \overline{0}$$

$$\Leftrightarrow (y, x) = 0$$

$$\Leftrightarrow y \perp x$$

$$\|x + y\|^2 = (x + y, x + y)$$

$$= (x, x) + (x, y) + (y, x) + (y, y)$$

$$= \|x\|^2 + \|y\|^2 \text{ by}$$

$$(x, y) = 0, (y, x) = 0 \tag{15}$$

$$\|x - y\|^2 = (x - y, x - y)$$

$$= (x, x) - (x, y) - (y, x) + (y, y)$$

$$= \|x\|^2 + \|y\|^2 \text{ by}$$

$$(x, y) = 0, (y, x) = 0 \tag{16}$$

From Eq. (15) and (16)

$$\|x + y\|^2 = \|x - y\|^2 = \|x\|^2 + \|y\|^2$$

4. Definition 3

A Vector x is said to be orthogonal to a non-empty set S (written as $x \perp s$) if $x \perp y \forall y \epsilon S$ [7].

4.1 Definition 3.1

The orthogonal complement of S denoted by $S^{\perp}$ is the set of all vectors orthogonal to s i.e., $S^{\perp} = \{x/x \in H \text{ and } x \perp s\}$.

i.e., $x \in S^{\perp} \Leftrightarrow x \perp s$ [10].

The following statements are the easy consequence of the definition

i. $\{0\}^{\perp} = H$

ii. $H^{\perp} = \{0\}$

iii. $S \cap S^{\perp} \subseteq \{0\}$

iv. $S_1 \subseteq S_2 \Rightarrow S_2^{\perp} \subseteq S_1^{\perp}$

v. $S^{\perp}$ is a closed subspace of H.

4.1.1 Theorem 3.1.1

If M is a proper closed linear subspace of a Hilbert Space H. Then there exists a non-zero vectors z_0 in H such that orthogonal to M. i.e., $z_0 \perp M$ [10].

4.1.1.1 Proof

Let x be a vector not in M and let d be the distance from r to M. Then by theorem "Let M be a closed linear subspace of a Hilbert Space H, let x be a vector not in M and let d be the distance from x to M . Then there exists a unique vector y_0 in M such that $\|x - y_0\| = d$"

We define $z_0 = x - y_0, x \in H, y_0 \in M$

$$\begin{aligned}
&\Rightarrow y_0 \in H \\
&\Rightarrow x - y_0 \in H \\
&\Rightarrow z_0 \in H \\
d = d(x, M) = &\inf\{\|x - m\| : m \in M\} \\
&\Rightarrow d \leq \|x - m\| \forall m \in M \text{ and} \\
&\|x - m\| \geq 0 \forall m \in M \\
&\inf\{\|x - m\| : m \in M\} \geq 0 \\
&\Rightarrow d \geq 0
\end{aligned} \tag{17}$$

Claim:- $d > 0$,

If $d = 0$ then $\inf\{\|x - m\| : m \in M\} = 0$.

Then there exists a sequence $\{m_n\}$ in M such that $\|x - m_n\| = 0$ as $n \rightarrow \infty$

$\Rightarrow \{m_n\} \rightarrow x$ as $n \rightarrow \infty$.

$\Rightarrow x \in M$, since M is closed.
This is a contradiction to $x \notin M$. $d \neq 0$.
Hence $d > 0$.

$$\|z_0\| = \|x - y_0\| = d$$
$$\Rightarrow \|z_0\| = d > 0$$
$$\Rightarrow \|z_0\| > 0$$
$$\Rightarrow z_0 \neq 0$$

$x \in M \Rightarrow x \in H$ since $M \subseteq H$

$$y_0 \in M \Rightarrow y_0 \in H$$
$$\Rightarrow x - y_0 \in H$$
$$\Rightarrow z_0 \in H$$

This proves the existence of non-zero vector z_0 in H .
We conclude the proof by showing that if y is an arbitrary vector in M then $z_0 \perp M$.
Let α be any scalar then

$$\|z_0 - \alpha y\| = \|x - y_0 - \alpha y\|$$
$$= \|x - (y_0 + \alpha y)\|$$
$$\geq \alpha \|z_0\|$$
$$\|z_0 - \alpha y\| \geq \|z_0\|$$
$$\Rightarrow (z_0 - \alpha y)(z_0 - \alpha y) \geq (z_0, z_0)$$
$$(z_0, z_0) - (z_0, \alpha y) - (\alpha y, z_0) + (\alpha y, \alpha y) \geq (z_0, z_0)$$
$$-\bar{\alpha}(z_0, y) - \alpha(y, z_0) + \alpha\bar{\alpha}(y, y) \geq 0 \tag{18}$$

Put $\alpha = \beta(z_0, y)$ for an arbitrary real β then $\bar{\alpha} = \beta(\overline{z_0, y})$ i.e. (18) becomes

$$\Rightarrow -\beta(\overline{z_0, y})(z_0, y) - \beta(z_0, y)(y, z_0) + \beta(z_0, y)\beta(\overline{z_0, y})\|y\|^2 = 0$$
$$\Rightarrow -\beta|z_0, y|^2 - \beta|(z_0, y)|^2 + \beta^2|(z_0, y)|^2\|y\|^2 \geq 0$$
$$\Rightarrow -2\beta|(z_0, y)|^2 + \beta^2|z_0, y|^2\|y\|^2 \geq 0$$
$$\Rightarrow \beta|(z_0, y)|^2\left(-2 + \beta\|y\|^2\right) \geq 0$$
$$\Rightarrow \beta|(z_0, y)|^2\left(\beta\|y\|^2 - 2\right) \geq 0 \tag{19}$$

Clearly

$$|(z_0, y)| = 0$$

Suppose

$$|(z_0, y)| > 0$$

Choose β arbitrary smallest +ve such that $\beta\|y\|^2 < 2$

$$\left(\beta\|y\|^2 - 2\right) < 0$$

$$\Rightarrow \beta|(z_0, y)|^2\left(\beta\|y\|^2 - 2\right) < 0$$

This is a contradiction to the Eq. (19)

$$|(z_0, y)| = 0$$

$$\Rightarrow (z_0, y) = 0$$

Therefore

$$z_0 \perp y, y \in M$$

$$\Rightarrow z_0 \perp M.$$

Hence it is proved.

5. Definition 4

5.1 Adjoint of an operator

Let H be a Hilbert Space and T be an operator on H then the mapping T^* : $H \to H$ defined by $(Tx, y) = (x, T^*y) \forall\, x, y \in H$ is called the adjoint of T . We verify that T^* is actually an operator on H [13]

i. To prove that T^* is linear

i.e., To prove

$T^*(y + z) = T^*(y) + T^*(z)$

$T^*(\alpha y) = \alpha T^*(y)$

$(x, T^*(y + z)) = (Tx, y + z)$

$= (x, T^*y) + (x, T^*z)$

$= (x, T^*y + T^*z)$

$(x, T^*(y + z)) = (x, T^*(y) + T^*(z))$

$\Rightarrow T^*(y + z) = T^*(y) + T^*(z)$

$(x, T^*(\alpha y)) = (Tx, \alpha y)$

$= \overline{\alpha}(Tx, y)$

$= \overline{\alpha}(x, T^*y)$

$(x, T^*(y + z)) = (x, \alpha T^*(y))$

$\Rightarrow T^*(\alpha y) = \alpha T^*(y)$

Therefore T is linear

ii. To prove that T^* is continuous

$0 \leq \|T^*(x)\|^2$

$= (T^*x, T^*x)$

$$= (TT^*x, x)$$
$$= |(TT^*x, x)|$$
$$\leq \|TT^*x\|\|x\|$$
$$\leq \|T\|\|T^*x\|\|x\| \text{ (by Schwartz Inequality)}$$
$$\|T^*x\| \leq \|T\|\|x\|$$
$$\sup\{\|T^*x\|/\|x\| \leq 1\} \leq \|T\| \text{ if } \|x\| \leq 1$$
$$\|T^*\| \leq \|T\|$$

Since T is bounded, T^* is also bounded.
Hence T^* is an operator on H .
These facts tell us that $T \to T^*$ is an mapping of $B(H)$ into itself
This mapping is called the adjoint operation on $B(H)$. [B(H) is a Banach Space].

5.2 Theorem 4.1

The adjoint operation $T \to T^*$ on $B(H)$ has the following properties [9]

i. $(T_1 + T_2)* = T_1^* + T_2^*$

ii. $(\alpha T)^* = \overline{\alpha} T^*$

iii. $(T_1 T_2)^* = T_2^* T_1^*$

iv. $T^{**} = T$

v. $\|T^*\| = \|T\|$

vi. $\|T^* T\| = \|T\|^2$

5.2.1 Proof

i. $(T_1 + T_2)^* = T_1^* + T_2^*$

$$(x, (T_1 + T_2)^* y) = ((T_1 + T_2)x, y)$$
$$= (T_1 x + T_2 x, y)$$
$$= (T_1 x, y) + (T_2 x, y)$$
$$= \left(x, T_1^* y\right) + \left(x, T_2^* y\right)$$
$$(x, (T_1 + T_2)^* y) = \left(x, T_1^* y\right) + \left(x, T_2^* y\right)$$
$$(T_1 + T_2)^* y = T_1^* y + T_2^* y$$
$$(T_1 + T_2)^* y = \left(T_1^* + T_2^*\right) y \forall y \in H$$
$$(T_1 + T_2)^* = T_1^* + T_2^*$$

ii. To prove $(\alpha T)^* = \overline{\alpha} T^*$

$$(x, (\alpha T)^* y) = ((\alpha T)x, y)$$

$$= (\alpha Tx, y)$$
$$= \alpha(Tx, y)$$
$$= \alpha(x, T^*y)$$
$$(x, (\alpha T)^*y) = \alpha(x, T^*y) \forall x \in H$$
$$(\alpha T)^* = \overline{\alpha} T^*$$

iii. To prove $(T_1T_2)^* = T_2^*T_1^*$

$$(x, (T_1T_2)^*y) = ((T_1T_2)x, y)$$
$$= (T_1(T_2x), y)$$
$$= (T_2x, T_1^*y)$$
$$(x, (T_1T_2)^*y) = (x, T_2^*T_1^*y)$$
$$(T_1T_2)^*y = T_2^*T_1^*y \forall y \in H$$
$$(T_1T_2)^* = T_2^*T_1^*$$

iv. To prove that $T^{**} = T$

$$(x, T^{**}y) = (T^*x, y) \forall x \in H$$
$$= \overline{(y, T^*x)}$$
$$= \overline{(Ty, x)} \forall x \in H$$
$$= (x, Ty)$$
$$T^{**}y = Ty$$
$$T^{**} = T$$

v. To prove that $\|T^*\| = \|T\|$

$$0 \leq \|T^*(x)\|^2$$
$$= (T^*x, T^*x)$$
$$= (TT^*x, x)$$
$$= |(TT^*x, x)|$$
$$\leq \|TT^*x\|\|x\|$$
$$\leq \|T\|\|T^*x\|\|x\| \text{ [by Schwartz Inequality].}$$
$$\|T^*x\| \leq \|T\|\|x\| \text{ if } \|x\| = 1].$$
$$\sup\{\|T^*x\|/\|x\| = 1\} \leq \|T\|$$
$$\|T^*x\| \leq \|T\| \tag{20}$$

We know that $T = T^{**}$.

$$\Rightarrow \|T\| = \|T^{**}\|$$
$$\leq \|T^*\|$$
$$\|T\| \leq \|T^*\| \tag{21}$$

From Eq. (20) and (21) we get

$$\|T^*\| = \|T\|$$

vi. To prove that $\|T^*T\| = \|T\|^2$.

$$0 \leq \|Tx\|^2$$
$$= (Tx, Tx)$$
$$= (Tx, T^{**}x)$$
$$= (T^*Tx, x)$$
$$= |(T^*Tx, x)|$$
$$\leq \|T^*Tx\|\|x\| \text{ by Schwartz Inequality]}$$
$$\leq \|T^*T\|\|x\|\|x\|$$
$$\|Tx\|^2 \leq \|TT^*\|\|x\|^2 \text{ if } \|x\| = 1$$
$$\|Tx\|^2 \leq \|TT^*\| \text{ if } \|x\| = 1$$
$$\sup\{\|Tx\|/\|x\| = 1\} \leq \|TT^*\|$$
$$\|T\|^2 \leq \|TT^*\| \tag{22}$$
$$\Rightarrow \|TT^*\| = \|T\|\|T^*\|$$
$$\leq \|T\|\|T\|$$
$$\leq \|T\|^2 \tag{23}$$

From Eq. (22) and (23) we get.

$$\|T^2\| = \|TT^*\|$$

6. Definition 5

6.1 Self Adjoint operator

An operator A on a Hilbert Space H is said to be self Adjoint if $A = A^*$. Since $0^* = 0$ and $I^* = I$, 0 and I are self-Adjoint operator [14].

6.2 Theorem 5.1

The Self-Adjoint operator in $B(H)$ from the closed real linear subspace of $B(H)$ and a real banch space which contains the identity transformation [14].

6.2.1 Proof

We will notice here about the product of two self-adjoint operators.
Let S denote the set of all Self-Adjoint operator in $B(H)$.
To prove $\underline{S}$ is a real linear subspace of $\underline{B(H)}$.
Let

$$A_1, A_2 \in S$$

$\Rightarrow A_1 \, and \, A_2$ are Self Adjoint.
$\Rightarrow A_1^* = A_1$ and $A_2^* = A_2$.
Let α, β are real,

$$(\alpha A_1 + \beta A_2)^* = (\alpha A_1)^* + (\beta A_2)^*$$
$$= \overline{\alpha} A_1^* + \overline{\beta} A_2^*$$
$$= \alpha A_1^* + \beta A_2^*$$
$$= \alpha A_1 + \beta A_2$$
$$(\alpha A_1 + \beta A_2)^* = \alpha A_1 + \beta A_2$$
$$\Rightarrow \alpha A_1 + \beta A_2 \in S$$

Therefore S is real linear subspace of $B(H)$.
Further if $\{A_n\}$ is a sequence of self Adjoint operators which converges to an operator A. Then it is easy to see that A is also self Adjoint.
i.e.) Let $\{A_n\}$ be a sequence in S such that $A_n \to A$.

$$\|A - A^*\| = \|A - A_n + A_n - A_n^* + A_n^* - A^*\|$$
$$\le \|A - A_n\| + \|A_n - A_n^*\| + \|A_n^* - A^*\|$$
$$\le \|A - A_n\| + \|A_n - A\| \text{ by } \|T^*\| = \|T\|.$$
$$\le 2\|A_n - A\| \to 0 \text{ as } A_n \to A$$

Therefore $\|A_n - A\| \le 0$

$$\Rightarrow \|A_n - A\| = 0 \text{ (since norm cannot be –ve)}$$
$$\Rightarrow A - A^* = 0$$
$$\Rightarrow A = A^*$$
$$\Rightarrow A \text{ is Self Adjoint}$$

Therefore $A \in S$.
Hence S is closed in $B(H)$.
Therefore $B(H)$ is complete, S is complete.
Hence S is a real Banach Space.
$\because I^* = I, I$ is self-Adjoint.
$\Rightarrow I \in S[(H)$ Contains identity transformation$]$.
Hence it is proved.

7. Conclusion

Even though every Hilbert Space is a Banach space, but there exist plenty of Banach space which are not Hilbert Spaces. However the converse is not true [13]. The Parallelogram Identity gives a criterion for normed space to become an inner product space [15]. It is important to emphasize that every finite dimensional normed Linear Space is a Hilbert Space [2]. Since every finite dimensional normed space is complete. The Theorems of this section allows us to define the adjoint Operator of a bounded Operator [3]. Finally we studied the Self adjoint Operator and its Properties. The results included here are Classical and can be found in the following Reference books in Functional Analysis.

Author details

Karthic Mohan[1*] and Jananeeswari Narayanamoorthy[2]

1 K.M.G. College of Arts and Science, Gudiyatham, Vellore District, Tamilnadu, India

2 Department of Mathematics, Arignar Anna Government Arts College for Women, Walajapet, Ranipet Dst, Tamilnadu, India

*Address all correspondence to: Karthicmohan2@gmail.com

References

[1] Simmons, GF. Introduction to Topology and Modern Analysis. New York: McGraw-Hill, 1963.

[2] Murphy G. C*-Algebras and Operator Theory. San Diego: Academic Press; 1990

[3] Young N. An Introduction to Hilbert Space. Cambridge, England: Cambridge University Press; 1988

[4] Zhu K. An Introduction to Operator Algebras. Boca Raton, USA: CRC Press; 1993

[5] Evan Jenkins B. Representation Theory, Symmetry, and Quantum Mechanics. Chicago, USA: University of Chicago; 2008

[6] Heinrich Saller L. The Basic Physical Lie Operations. Munich, Germany: Max-Planck Institute for Physics; 2004

[7] John DePree S. Introduction to Real Analysis. New York: John Wiley and Sons; 1988

[8] Joseph Maciejko K. Representations of Lorentz and Poincare Groups. Stanford, California: MITP and Department of Physics, Stanford University; 2012

[9] Kreyszig E, Birkho G. The establishment of functional analysis. Historia Mathematica. 1984;**11**: 258-321

[10] Joachim Weidmann M. Linear Operators in Hilbert Space. New York, USA: Springer Karstedt; 1980

[11] Conway J. A Course in Functional Analysis. New York: Springer-Verlag; 1990

[12] James R, editor. Topology (Second Edition). Mungres Eastern Economy Edition. New York, USA: Pearson; 2018

[13] Kreysig E. Introductory Functional Analysis with Applications. New York, USA: John Wiley and Sons; 1978

[14] Debnath L, Mikusinskin P. Introduction to Hilbert Space with Applications. New York, USA: John Wiley and Sons; 2005

[15] Rudvin W. Functional Analysis. New York: MacGraw-Hill; 1973

Chapter 2

Operator Topology for Logarithmic Infinitesimal Generators

Yoritaka Iwata

Abstract

Generally unbounded infinitesimal generators are studied in the context of operator topology. Beginning with the definition of seminorm, the concept of locally convex topological vector space is introduced as well as the concept of Fréchet space. These are the basis for defining operator topologies. Consequently, by associating the topological properties with the convergence of sequence, a suitable mathematical framework for obtaining the logarithmic representation of infinitesimal generators is presented.

Keywords: operator theory, locally strong topology, infinitesimal generator

1. Introduction

Let X be an infinite/finite dimensional Banach space with the norm $\|\cdot\|$, and Y be a dense subspace of X. The Cauchy problem for abstract evolution equation of hyperbolic type [1, 2] is defined by

$$\begin{aligned} &du(t)/dt - A(t)u(t) = f(t), \qquad t \in [0, T], \\ &u(0) = u_0 \end{aligned} \tag{1}$$

in X, where $A(t)$: $Y \to X$ is assumed to be the infinitesimal generator of evolution operator $U(t,s)$ satisfying the strong continuity (for the definition of strong topology, refer to the following section) and the semigroup property:

$$U(t,s) = U(t,r)U(r,s) \tag{2}$$

for $0 \le s \le r \le t < T$. $U(t,s)$ is a two-parameter C_0-semigroup of operator that is a generalization of one-parameter C_0-semigroup and therefore an abstract generalization of the exponential function of operator. For an infinitesimal generator $A(t)$ of $U(t,s)$, the solution $u(t)$ is represented by $u(t) = U(t,s)u_s$ with $u_s \in X$ for a certain $0 \le s \le T$ (cf. Hille-Yosida Theorem; for example see [3–5]).

2. Operator topology

2.1 The dual formalism of evolution equation

The dual space of X being denoted by X^* is defined by

$$X^* = L(X,K), \tag{3}$$

where K is a scalar field making up the space X, and $L(X,K)$ denotes the space of continuous linear functionals. Since K is also a Banach space, $L(X,K)$ satisfies the properties of Banach space.

Let $\langle\cdot,\cdot\rangle: X \times X^* \to \mathbb{C}$ be a dual product between X and X^*, and $\mathbb{C}$ be a set of complex numbers. The adjoint operator $A(t)^*: D(A(t)) \to X^*$ is defined by the operator satisfying

$$\langle A(t)u, v\rangle = \langle u, A(t)^* v\rangle \tag{4}$$

for any $u \in D(A(t))$ and $v \in D(A(t)^*)$. If X is a Hilbert space, the dual product is replaced with a scalar product $(\cdot,\cdot)$ equipped with X. Unique dual correspondence is valid if X is strictly convex Banach space at least (for convex Banach space, see [6]). By taking the dual product, the abstract evolution equation in X:

$$du(t)/dt - A(t)u(t) = f(t), \qquad t \in [0,T] \tag{5}$$

is written as a scalar-valued evolution equation in $\mathbb{C}$:

$$\langle du(t)/dt, v(t)\rangle - \langle A(t)u(t), v(t)\rangle = \langle f(t), v(t)\rangle, \qquad t \in [0,T] \tag{6}$$

for a certain $v(t) \in X^*$. The formalism (6), which is associated with the Gelfand triplets [7], has been considered by variational method of abstract evolution equations [8, 9]. Eqs. (5) and (6) cannot necessarily be equivalent in the sense of operator topology.

2.2 Locally strong topology

Let $p(u)$ be a seminorm equipped with a space $\mathcal{X}$, and the family of seminorms be denoted by P. Locally convex spaces are the generalization of normed spaces. Here the topology is called locally convex, if the topology admits a local base at 0 consisting of balanced, absorbent, and convex sets. In other words, a topological space $\mathcal{X}$ is called locally convex, if its topology is generated by a family of seminorms satisfying

$$\cap_{p \in P}\{u \in \mathcal{X}; p(u) = 0\} = 0_{\mathcal{X}}, \tag{7}$$

where $0_{\mathcal{X}}$ denotes the zero of topological space $\mathcal{X}$. Fréchet spaces are locally convex spaces that are completely metrizable with a certain complete metric. It follows that a Banach space X is trivially a Fréchet space.

For Banach spaces X and Y, the bounded linear operators from X to Y is denoted by $B(X,Y)$. In particular $B(X,X)$ is written by $B(X)$. The operator space $B(X)$ is called the Banach algebra, since it holds the structure of algebraic ring. The norm of $B(X)$, which means the operator norm, is defined by

$$\|T\|_{B(X)} = \sup_{x \neq 0} \frac{\|Tx\|_X}{\|x\|_X}. \tag{8}$$

A norm is trivially a seminorm. Consequently, the topological space in this article is set to be a Banach space $B(X)$.

There are several standard typologies defined on $B(X)$. The topologies listed below are all locally convex, which implies that they are defined by a family of seminorms. The topologies are identified by the convergence arguments. Let $\{T_n\}$ be a sequence in a Banach space X.

- $T_n \to T$ in the uniform topology, if $\|T_n - T\|_{B(X)} \to 0$;
- $T_n \to T$ in the strong topology, if $T_n x \to Tx$ for any $x \in X$;
- $T_n \to T$ in the weak topology, if $F(T_n x) \to F(Tx)$ for any $F \in X^*$ and $x \in X$;

where the uniform topology is the strongest, and the weak topology is the weakest. Indeed a topology is called stronger if it has more open sets and weaker if it has less open sets. If Y is a vector space of linear maps on the vector space X, then a topology $\sigma(X, Y)$ is defined to be the weakest topology on X such that all elements of Y are continuous. The topology of $\sigma(X, Y)$ type is apparent if the formalism (6) is considered; the weak topology is written by $\sigma(B(X), B(X)^*)$. Although there are some intermediate topologies between the above three; strong* topology, weak* topology, and so on, another type of topology is newly introduced in this article.

Definition 1 (locally strong topology)

- $T_n \to T$ in the locally strong topology, if $T_n \overline{x} \to T\overline{x}$ for a certain $\overline{x} \in X$.

This topology is utilized to define a weak differential appearing in the logarithmic representation of infinitesimal generators.

3. Infinitesimal generator

3.1. Logarithmic infinitesimal generator

The logarithm of evolution operator is represented using the Riesz-Dunford integral. A time interval $[0, T]$ with $0 \le s, t \le T$ is provided. For a certain $u_s \in X$, let a trajectory $u(t) = U(t,s)u_s$ be given in a Banach space X. For a given $U(t,s) \in B(X)$, its logarithm is well defined [10]; there exists a certain complex number κ satisfying

$$\mathrm{Log}(U(t,s) + \kappa I) = \frac{1}{2\pi i} \int_\Gamma \mathrm{Log}\lambda \, (\lambda - \kappa - U(t,s))^{-1} d\lambda, \tag{9}$$

where an integral path Γ, which excludes the origin, is a circle in the resolvent set of $U(t,s) + \kappa I$.

Let us call $\mathrm{Log}(U(t,s) + \kappa I)$ the alternative infinitesimal generator to $A(t)$. Since the alternative infinitesimal generator [11]

$$a(t,s) := \mathrm{Log}(U(t,s) + \kappa I) \tag{10}$$

is necessarily bounded on X, its exponential function $e^{a(t,s)}$ is always well defined as a convergent power series. Note that the alternative infinitesimal generator $a(t,s)$ is bounded on X, although the corresponding infinitesimal generator $A(t)$ is possibly an unbounded operator. It follows that $e^{-a(t,s)} = \left(e^{a(t,s)}\right)^{-1}$ is automatically well defined if $e^{a(t,s)}$ is well defined. Also $e^{a(t,s)}$ is invertible regardless of the validity of the invertible property for original $U(t,s)$. The logarithmic representation of infinitesimal generator (logarithmic infinitesimal generator, for short) is obtained as follows.

Lemma 1 (Logarithmic infinitesimal generators [10]). Let t and s satisfy $0 \le t, s \le T$, and Y be a dense subspace of X. If $A(t)$ and $U(t,s)$ commute, infinitesimal generators $\{A(t)\}_{0 \le t \le T}$ are represented by means of the logarithm function; there exists a certain complex number $\kappa \neq 0$ such that

$$A(t)\,u = \left(I - \kappa e^{-a(t,s)}\right)^{-1} \partial_t a(t,s)\,u, \tag{11}$$

where u is an element of a dense subspace Y of X, and ∂_t is a kind weak differential being defined by the locally strong topology.

Proof. Only formal discussion is made here (for the detail, see [10, 12]).

$$(U(t,s) + \kappa I)\partial_t a(t,s) = (U(t,s) + \kappa I)(U(t,s) + \kappa I)^{-1}\partial_t U(t,s). \tag{12}$$

It leads to

$$\begin{aligned} A(t)\,u &= U(t,s)^{-1}\partial_t U(t,s)u = U(t,s)^{-1}(U(t,s) + \kappa I)\partial_t a(t,s)\,u \\ &= \left(I + \kappa\left(e^{a(t,s)} - \kappa I\right)^{-1}\right)\partial_t a(t,s)\,u \\ &= \left(e^{a(t,s)} - \kappa I + \kappa I\right)\left(e^{a(t,s)} - \kappa I\right)^{-1} \partial_t a(t,s)\,u \\ &= \left(I - \kappa e^{-a(t,s)}\right)^{-1} \partial_t a(t,s)\,u, \end{aligned} \tag{13}$$

where u is an element in Y. □

The solution trajectory is given, and $\{A(t)\}_{0 \le t \le T}$ are determined for one fixed $u \in Y$. Here is a reason why the locally-strong topology is introduced. Eq. (11) is the logarithmic representation of infinitesimal generator $A(t)$. This representation is useful not only to mathematical analysis but also to operator algebra [12, 13].

3.2. Differential operator in the logarithmic representation

The convergence of the limit in the differential operator ∂_t of Eq. (11) is discussed. The convergence in the locally strong topology is applied to the evolution operator $U(t,s) \in B(X)$.

Definition 2 (Weak limit using the locally strong topology). For $0 \le t, s \le T$, the weak limit

$$\lim_{h\to 0} h^{-1}(U(t+h,s) - U(t,s))\,u_s = \lim_{h\to 0} h^{-1}(U(t+h,t) - I)\,U(t,s)\,u_s \tag{14}$$

is assumed to exist for a certain u_s in a dense subspace Y of X. The limit " lim " is practically denoted by "w*lim* " in the following, since it is a limit in a kind of weak topology.

Let t-differential of $U(t,s)$ in a weak sense of the above be denoted by ∂_t, then it follows that

$$\partial_t U(t,s)\,u_s = A(t)U(t,s)\,u_s, \tag{15}$$

and a generalized concept of infinitesimal generator $A(t) : Y \to X$ is introduced by

$$A(t)u_s := \operatorname{wlim}_{h\to 0} h^{-1}(U(t+h,t) - I)u_s \tag{16}$$

for a certain $u_s \in Y$, where the convergence in w*lim* must be replaced with the strong convergence in the standard theory of abstract evolution equations [4].

The operator $A(t)$ defined in this way for a whole family $\{U(t,s)\}_{0 \le t,s \le T}$ is called the pre-infinitesimal generator in [10], because only its exponetiability with a certain ideal domain is ensured without justifying the dense property of its domain space. Indeed pre-infinitesimal generators are not necessarily infinitesimal generators, while infinitesimal generators are pre-infinitesimal generators.

4. Main result

According to the standard theory of abstract evolution equation [4], the evolution operator is assumed to be strongly continuous. It follows that the trajectory $U(t,s)u_s$ is continuous in X. Here is the reason why it is sufficient to consider the convergence of differential operator ∂_t only with a fixed element $\overline{u} = u_s \in Y \subset X$ with $0 \leq s \leq T$. Also, in terms of analyzing the trajectory in finite-/infinite-dimensional dynamical systems, it is reasonable to consider the convergence in the topology uniquely sticking to the trajectory. Consequently the infinitesimal generator can be extracted by one sample point in the interval (**Figure 1**). Indeed, according to the independence between t and s,

$$A(t)u_s = \operatorname*{wlim}_{h \to 0} h^{-1}(U(t+h,t) - I)u_s \tag{17}$$

is true for any $t \in [s, T]$, once $A(t)$ is obtained for a sample point $u_s \in Y$. Such a restrictive topological treatment contributes to generalize or weaken the differential.

For a given evolution operator $U(t,s) \in B(X)$, the profile of locally strong topology is obtained in this article. In Banach space $B(X)$, a subset $F \subset B(X)$ is a closed set, if and only if

$$\{a_n\} \in F, a \in B(X), a_n \to a \Rightarrow a \in F \tag{18}$$

is satisfied ($n = 1, 2, \cdots$), where the operation of limit depends on a chosen topology. Here the following two theorems are proved to clarify the mathematical property of the locally strong topology.

Theorem 1.1. The locally strong topology is weaker than the strong topology.

Proof. It is enough to prove that a closed set in strong topology is closed in the locally strong topology. Let an arbitrary closed set of $B(X)$ in the strong topology be V. It satisfies

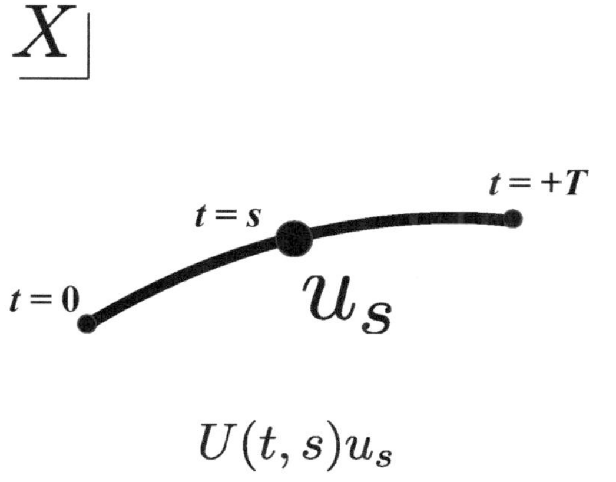

Figure 1.
Trajectory $U(t,s)u_s$ in X. $U(t,s) \in B(X)$ is assumed to be strongly continuous with respect to time variables, so that a trajectory $U(t,s)u_s$ is continuous in X. Note that it is necessary to replace the trajectory $U(t,s)u_s$ with the regularized trajectory $(e^{a(t,s)} - \kappa I)u_s$ to consider the negative time evolution [11, 12].

$$\{T_n\} \in V, T \in B(X), \lim_{n\to\infty} \|T_n x - Tx\| = 0 \quad \Rightarrow \quad T \in V \tag{19}$$

for an arbitrary $x \in X$. In the locally strong topology ($\|(T_n - T)\bar{x}\|$ for a certain $\bar{x} \in X$), the convergence $T_n \to T \in V$ is true. □

Theorem 1.2. The locally strong topology is not necessarily stronger than the weak topology.

Proof. The proof is carried out in the similar manner to Theorem 1.1. Let an arbitrary closed set of $B(X)$ in the locally strong topology be V. It satisfies

$$\{T_n\} \in V, T \in B(X), \lim_{n\to\infty} \|T_n \bar{x} - T\bar{x}\| = 0 \quad \Rightarrow \quad T \in V \tag{20}$$

for a fixed $\bar{x} \in X$. By taking the dual product of an arbitrary $F \in X^*$, it follows that

$$\{T_n\} \in V, a \in B(X), \lim_{n\to\infty} \langle (T_n - T)\bar{x}, F \rangle = 0 \quad \Rightarrow \quad T \in V \tag{21}$$

It shows that the closedness of V in a locally weak topology, where the locally weak topology is defined by fixing the weak topology with $x = \bar{x}$ in the same manner as the locally strong topology.

On the other hand, weak convergence cannot be assured if $x \neq \bar{x}$. Indeed, for $x_1 \neq \bar{x}$, the statement

$$\{T_n\} \in V, a \in B(X), \lim_{n\to\infty} \langle (T_n - T)x_1, F \rangle = 0 \quad \Rightarrow \quad T \in V \tag{22}$$

does not follow from the statement (20). It shows that there is no guarantee for locally strong topology to be stronger than the weak topology. □

5. Summary

The concept of locally strong topology is introduced by the proofs clarifying its specific topological weakness. The locally strong topology is a topology unique to the solution trajectory of abstract evolution equations (**Figure 1**). That is, the locally strong topology holds the one-dimensionality specific to a certain trajectory. Although the locally strong topology has already been utilized even without the nomenclature to clarify the algebraic structure of semigroups of operators and their infinitesimal generators, those fundamentals are made in this article. The locally strong topology is also expected to be useful to analyze each single trajectory defined in finite-/infinite-dimensional dynamical systems.

Acknowledgements

The author is grateful to Prof. Emeritus Hiroki Tanabe for valuable comments. This work was partially supported by JSPS KAKENHI Grant No. 17K05440.

Author details

Yoritaka Iwata
Faculty of Chemistry, Materials and Bioengineering, Kansai University, Osaka, Japan

*Address all correspondence to: iwata_phys@08.alumni.u-tokyo.ac.jp

References

[1] Kato T. Linear evolution equation of "hyperbolic" type. Journal of the Faculty of Science, University of Tokyo. 1970; **17**:241-258

[2] Kato T. Linear evolution equation of "hyperbolic" type II. Journal of the Mathematical Society of Japan. 1973; **25**(4):648-666

[3] Kato T. Perturbation Theory for Linear Operators. Berlin, Heidelberg, New York: Springer-Verlag; 1966

[4] Tanabe H. Equations of Evolution. Pitman; 1979

[5] Yosida K. Functional Analysis. Berlin, Heidelberg: Springer-Verlag; 1965

[6] Brezis H. Functional Analysis, Sobolev Spaces and Partial Differential Equations. New York: Springer-Verlag; 2010

[7] Gelfand M, Vilenkin NY. Generalized Functions. Vol. 4. Applications of Harmonic Analysis. Academic Press; 1964

[8] Lions JL. Equations Differentielles Operationnelles et Problemes aux Limits. Berlin, Heidelberg: Springer-Verlag; 1961

[9] Lions JL, Magenes E. Non-homogeneous Boundary Value Problems and Applications. Berlin, Heidelberg: Springer-Verlag; 1972

[10] Iwata Y. Infinitesimal generators of invertible evolution families. Methods of Functional Analysis and Topology. 2017;**23**(1):26-36

[11] Iwata Y. Alternative infinitesimal generator of invertible evolution families. Journal of Applied Mathematics and Physics. 2017;**5**: 822-830

[12] Iwata Y. Theory of $B(X)$-Module. arXiv:1907.08767

[13] Iwata Y. Operator algebra as an application of logarithmic representation of infinitesimal generators. Journal of Physics: Conference Series. 2018;**965**:012022

Chapter 3

New Topology on Symmetrized Omega Algebra

Mesfer Alqahtani, Cenap Özel and Ibtesam Alshammari

Abstract

The purpose of this paper is to define a new topology called symmetrized omega algebra topology and discuss some of its topological properties. Two different examples from an ordered infinite set of symmetrized omega topology are introduced. Furthermore, we study the relationship between symmetrized omega topology and weaker kinds of normality.

Keywords: tropical geometry, idempotent semiring, topological space, topological properties, omega algebra and symmetrized omega algebra

1. Introduction

Tropical geometry is the most recent but fast growing branch of mathematical sciences, which is analytically based on idempotent analysis and algebraically on idempotent semirings also known as tropical semirings. These are basically extended sets of real numbers $\mathbb{R}_\infty : \mathbb{R} \cup \{\infty\}$ and $\mathbb{R}_{-\infty} : \mathbb{R} \cup \{-\infty\}$ which are given monoidal structures by using min and max operations for addition, respectively. In order to adhere to the semiring structure, the additive operation of $\mathbb{R}$ is used as the multiplication operation. By these choices, both $\mathbb{R}_\infty$ and $\mathbb{R}_{-\infty}$ become idempotent semirings. The literature, they are also termed as min and max plus algebras, respectively. In both cases, 0 of $\mathbb{R}$ becomes a multiplicative identity and ∞ and $-\infty$ become additive identities of these semirings, respectively. Interestingly, some authors associate $\mathbb{R}_{-\infty}$ to tropical geometry, while other authors associate $\mathbb{R}_\infty$ to tropical geometry (see [1–4]). Omega algebra or "$\omega-$ algebra" for short, unifies the different terms and introduces an original structure, which, in fact, is an "abstract tropical algebra". The $\mathbb{R}_{-\infty}$ and $\mathbb{R}_\infty$ and their nearby structures, like $\min - \max$ and $\max -$ times algebras, etc., are all subsumed under omega algebra. All these are idempotent semirings, which are also called dioids. In previous studies, for the construction of all such semirings, an ordered infinite abelian group is mandatory. In $\omega-$ algebra, the definition is extended to cyclically ordered abelian groups and also to finite sets under some suitable ordering. Note that cyclically ordered abelian groups are more general than that of ordered abelian groups [5]. The aim of this paper is to define a new topology on symmetrized omega algebra, and discus some of its topological properties. Two different examples from an ordered infinite sets of symmetrized omega topology are introduced. Furthermore, we study the relationship between symmetrized omega topology and weaker kinds of normality. Our paper is organized as follows. In Section 2, we review an abstract definition for some basic facts about abstract omega algebras. In addition, we give a brief of symmetrized omega algebra and rules of calculation in omega. In Section 3, we define a new topology on symmetrized omega algebra and discuss some of its topological properties. In Section 4, we provide two different examples of symmetrized omega topology: the first and second examples are from an ordered infinite set. Finally, we study the relationship between symmetrized omega topology and weaker kinds of normality in Section 5. Throughout this paper, we do not assume T_2 in the definition of compactness. We also do not assume regularity in the definition of Lindelöfness.

The ideas from this paper were taken from the PhD thesis of Mr. Mesfer Hayyan Alqahtani in King Abdulaziz University.

2. Preliminaries

In this section, we provide an abstract definition for review some basic facts about abstract omega algebra. Furthermore, we also provide a brief of symmetrized omega algebra and rules of calculation in omega. For more details, see [6].

2.1 Omega algebra

Let $(G, \circ, e)$ be an abelian group. Let A be a closed subset of G and $e \in A$. Then $(A, \circ, e)$ is a submonoid of G. Assume that ω is an indeterminate (may belong to A or G, as we will see in Examples 1 and 2). Obviously, in this case ω is no longer an indeterminate. Because the terms are generated from tropical geometry, this indeterminate can be called a tropical indeterminate.

Definition 1. [6]

We say that $A_\omega = A \cup \{\omega\}$ is an *omega algebra* (in short $\omega-$ *algebra*) over the group G in case A_ω is closed under two binary operations,

$$\oplus, \otimes : A_\omega \times A_\omega \to A_\omega, \tag{1}$$

such that $\forall a_1, a_2, a_3 \in A$, the following axioms are satisfied:

i. $a_1 \oplus a_2 = a_1 \; or \; a_2$;

ii. $a_1 \oplus \omega = a_1 = \omega \oplus a_1$;

iii. $\omega \oplus \omega = \omega$;

iv. $a_1 \otimes a_2 = a_2 \otimes a_1 \in A$;

v. $(a_1 \otimes a_2) \otimes a_3 = a_1 \otimes (a_2 \otimes a_3)$;

vi. $a_1 \otimes e = a_1$;

vii. $a_1 \otimes \omega = \omega \otimes a_1 = \begin{cases} \omega & if \quad \omega \neq e \\ a_1 & if \quad \omega = e \end{cases}$;

viii. $\omega \otimes \omega = \omega$;

ix. $a_1 \otimes (a_2 \oplus a_3) = (a_1 \otimes a_2) \oplus (a_1 \otimes a_3)$.

Remark 2. [6]

1 $\oplus$ is a pairwise comparison operation such as max, min, inf, sup, up, down, lexicographic ordering, or anything else that compairs two elements of A_ω.

Obviously, it is associative and commutative and the tropical indeterminate ω plays the role of the identity. Hence $(A_\omega, \oplus, \omega)$ is a commutative monoid.

2. $\otimes$ is also associative and commutative on A_ω, and e plays the role of the multiplicative identity of A_ω. Hence, $(A_\omega, \otimes, e)$ is also a commutative monoid.

3. The left distributive law (ix) also gives the right distributive law.

4. Every element of A_ω is an idempotent under $\oplus$.

5. Altogether, we write both structures as: $A_\omega = (A_\omega, \oplus, \otimes, \omega, e)$. This is an idempotent semiring.

Remark 3. [6] A $\omega-$ algebra can similarly be defined over a commutative monoid, ring, or even a semiring. More generally, one may construct analogously such algebras on other weaker structures. In this note, we confined ourselves to only $\omega-$ algebras over abelian groups and rings.

Example 4. [6] *Max-plus algebra, min-plus algebra* and all such "so called" *algebras* are particular cases of the $\omega-$ *algebra* over the ring $\mathbb{R}$ or its associated subrings. A simpler example is the following. In the abelian group $(\mathbb{Z}, +)$, for any integer m, we have $W(m) = \{0, m, 2m, \cdots\}$. This is an additive submonoid of $(\mathbb{Z}, +)$. Let $\omega = -\infty$, $a_1 \oplus a_2 = \max(a_1, a_2)$ and $a_1 \otimes a_2 = a_1 + a_2, \forall a_1, a_2 \in W(m)$. Then,

$$W(m)_{-\infty} = \left(W(m)_{-\infty}, \oplus, \otimes, -\infty, 0\right) \tag{2}$$

is $-\infty$ $-$ algebra over the abelian group of integers $\mathbb{Z}$. Hence, we have a sequence of $\omega-$ subalgebras

$$W(m) \geq W(2m) \geq \cdots.$$

Example 5. [6] Cartesian products of omega algebras. In this example, we explain a construction of an omega algebra from other given omega algebras. Let $\{(G_i, \circ_i, e_i) : i = 1, \cdots, n\}$ be abelian groups and $\{(A_{\omega_i}, \oplus_i, \otimes_i, \omega_i, e_i) : i = 1, \cdots, n\}$ be a respective family of omega algebras, where ω_i are tropical indeterminate. As usual, we define the Cartesian product as

$$\mathcal{X}_\omega = A_{\omega_1} \times \cdots \times A_{\omega_n} = \{(a_1, \cdots a_n) : a_i \in A_{\omega_i}; i = 1, \cdots, n\}. \tag{3}$$

In order to provide a convenient technique to give an additive structure to $\mathcal{X}_\omega$, we assume that the $n-tuples$ $\mathbf{a} = (a_1, \cdots a_n)$, $\mathbf{b} = (b_1, \cdots b_n) \in \mathcal{X}_\omega$ are in lexicographic ordering. Then, to define the sum

$$\mathbf{a} \oplus \mathbf{b} = \mathbf{a} \text{ or } \mathbf{b} \tag{4}$$

by using the following rules:

$$\text{If } a_1 \oplus_1 b_1 = a_1 \text{ then } \mathbf{a} \oplus \mathbf{b} = \mathbf{a}. \tag{5}$$

$$\text{If } a_i = b_i \text{ for } 1 \leq i \leq k \leq n, \text{ and } a_{k+1} \oplus_{k+1} b_{k+1} = a_{k+1}, \text{ then, } \mathbf{a} \oplus \mathbf{b} = \mathbf{a}. \tag{6}$$

Similarly, rules for $\mathbf{a} \oplus \mathbf{b} = \mathbf{b}$ can be determined. Multiplication can be to define component wise. Thus,

$$\mathbf{a} \otimes \mathbf{b} = (a_1 \otimes_1 b_1, \cdots, a_n \otimes_n b_n). \tag{7}$$

The other rules of the Definition 1 can straightforwardly be verified. Hence, $(\mathcal{X}_\omega, \oplus, \otimes, \omega, e)$, where $\omega = (\omega_1, \cdots, \omega_n)$ is the additive identity and $e = (e_1, \cdots, e_n)$ is the multiplicative identity of $\mathcal{X}_\omega$, is an omega algebra over the Cartesian product of abelian groups $G_1 \times \cdots \times G_n$..

2.2 The symmetrized omega algebra

Let $(G, \circ, e)$ be an abelian group and $(A_\omega, \oplus, \otimes, \omega, e)$ an $\omega-$algebra over the group G. Following the method used in constructing integers from the natural numbers, we consider the set of ordered pairs $\mathcal{P}_\omega = A_\omega^2$ with component wise addition $\oplus$, for all (a,b), $(c,d) \in \mathcal{P}_\omega$,

$$(a,b) \oplus (c,d) = (a \oplus c, b \oplus d) \tag{8}$$

Because of the four possibilities (a,b), (a,d), (c,d) or (c,b) for the result, the addition in (8), is ambiguous. As our goal from constructing the algebra of pairs is the construction of the symmetrized omega algebra of A_ω, we are in front of two possibilities: One is to use –for $n = 2$, and define an equivalence relation $\sim$ on the $\omega-$algebra of pairs which is compatible with relevant operations, and the other is to define an equivalence relation on the set $\mathcal{P}_\omega$ that allows the component wise addition to be defined in the quotient set.

First Construction, let $\leq$ be the ordering defined on A_ω by the relation

$$a \leq b \Leftrightarrow a \oplus b = b \tag{9}$$

which gives a total order on A_ω and for all $a \in A_\omega$, we have $\omega \leq a$. For $a \neq b$, such that $a \oplus b = b$, we denote by $a < b$. Under the ordering $\leq$, rules (5) and (6) defined in Example 5, are satisfied on $\mathcal{P}_\omega = A_\omega^2$ and so $\mathcal{P}_\omega$ is an $\omega-$algebra under the addition defined in 1 and the component wise multiplication. Let ∇ be the relation defined on $\mathcal{P}_\omega$ as follows: for all (a,b), $(c,d) \in \mathcal{P}_\omega$

$$(a,b) \nabla (c,d) \Leftrightarrow a \oplus d = b \oplus c. \tag{10}$$

Then ∇ is reflexive and symmetric but not transitive for A_ω contains more than 4 elements. In fact, let $a, b, c, d \in A_\omega$ such that $a < b < c < d$, then we have

$$a \oplus d = d = b \oplus d = c \oplus d \text{ and } a \oplus c = c \neq b = b \oplus b$$

which give $(a,b) \nabla (d,d)$ and $(d,d) \nabla (b,c)$, but there is no relation between (a,b) and (b,c). As ∇ is not an equivalence relation, we cannot use it to obtain the quotient $\omega-$algebra $\frac{\mathcal{P}_\omega}{\nabla}$ (like the one to obtain integers from the natural numbers).

Definition 6. [6] Let $\sim$ be the equivalence relation close to ∇ defined as follows: for all (a,b), $(c,d) \in \mathcal{P}_\omega$,

$$(a,b) \sim (c,d) \Leftrightarrow \begin{cases} (a,b) \nabla (c,d) & \text{if } a \neq b \text{ and } c \neq d \\ (a,b) = (c,d) & \text{otherwise} \end{cases} \tag{11}$$

In addition to the class element $\overline{\omega} = \overline{(\omega, \omega)}$; for all $a \in A_\omega$, with $a \neq \omega$, we have three kinds of equivalence classes:

1. $\overline{(a, \omega)} = \{(a,b) \in \mathcal{P}_\omega, b < a\}$, called positive $\omega-$element.

2. $\overline{(\omega, a)} = \{(b, a) \in \mathcal{P}_\omega, b < a\}$, called negative $\omega-$element.

3. $\overline{(a, a)}$ called balanced $\omega-$element.

Unfortunately, the addition defined by (7) and rules (8) and (9) in Example 5 is not compatible with the equivalence relation in $\mathcal{P}_\omega$, because for (a, ω), (a, b), (ω, c), $(d, c) \in \mathcal{P}_\omega$, such that

$$\begin{cases} (a, \omega) \sim (a, b) \\ (\omega, c) \sim (d, c), \end{cases} \tag{12}$$

we have

$$(a, \omega) \oplus (\omega, c) \sim (a, b) \oplus (d, c) \text{ iff } (a, b) \oplus (d, c) = (a, b) \tag{13}$$

$$\text{and if } (a, b) \oplus (d, c) = (d, c), \tag{14}$$

then there is no compatibility. So the omega algebra of pairs cannot produce the symmetrized omega algebra.

Second Construction

Proposition 7. [6]

The addition operation $\overline{\oplus}$ defined by

$$\overline{(a, b)}\,\overline{\oplus}\,\overline{(c, d)} = \overline{(a \oplus c, b \oplus d)}$$

on the quotient set $\frac{\mathcal{P}_\omega}{\sim}$ is well defined and satisfies the axioms (i), (ii) and (iii) of Definition 1, with the zero class element $\overline{\omega} = \overline{(\omega, \omega)}$, except this case $\overline{(a, \omega)}\,\overline{\oplus}\,\overline{(\omega, a)} = \overline{(\omega, a)}\,\overline{\oplus}\,\overline{(a, \omega)} = \overline{(a, a)}$, where $a \in A_\omega \backslash \{\omega\}$ does not satisfy the axiom (i).

Proposition 8. [6]

i. The set $\frac{\mathcal{P}_\omega}{\sim}$ is closed under the binary multiplication operation $\overline{\otimes}$ defined as follows: for all $\overline{(a, b)}$, $\overline{(c, d)} \in \frac{\mathcal{P}_\omega}{\sim}$;

$$\overline{(a, b)}\,\overline{\otimes}\,\overline{(c, d)} = \overline{((a \otimes c) \oplus (b \otimes d), (a \otimes d) \oplus (b \otimes c))} \tag{15}$$

and satisfies axioms from (iv) to (ix) of Definition 1, with the unit class element $\overline{e} = \overline{(e, \omega)}$.

ii. In addition, we have for all $a, b \in A_\omega$

a. $\overline{(a, \omega)}\,\overline{\otimes}\,\overline{(b, \omega)} = \overline{(a \otimes b, \omega)}$;

b. $\overline{(a, \omega)}\,\overline{\otimes}\,\overline{(\omega, b)} = \overline{(\omega, a \otimes b)}$;

c. $\overline{(a, \omega)}\,\overline{\otimes}\,\overline{(b, b)} = \overline{(a \otimes b, a \otimes b)}$;

d. $\overline{(\omega, a)}\,\overline{\otimes}\,\overline{(b, b)} = \overline{(a \otimes b, a \otimes b)}$.

Definition 9. [6] The structure $\left(\frac{\mathcal{P}_\omega}{\sim}, \overline{\oplus}, \overline{\otimes}, \overline{\omega}, \overline{e}\right)$ is called the symmetrized $\omega-$algebra over the abelian group $G \times G$ and we denote it by $\mathbb{S}_\omega$.

In the coming sections just for simplicity we will only use $\oplus$ and $\otimes$ instead the operations $\overline{\oplus}$ and $\overline{\otimes}$, respectively.

Remark 10. [6]

1. Despite the nature of the positive and the negative $\omega-$elements, they are not the inverses of each other for the additive operation $\overline{\oplus}$,

2. We have three symmetrized $\omega-$subalgebras of $\mathbb{S}_\omega$,

$$\mathbb{S}_\omega^{(+)} = \left\{\overline{(a,\omega)}, a \in A_\omega\right\},$$
$$\mathbb{S}_\omega^{(-)} = \left\{\overline{(\omega,a)}, a \in A_\omega\right\},$$
$$\mathbb{S}_\omega^{(0)} = \left\{\overline{(a,a)}, a \in A_\omega\right\}.$$

3. The three symmetrized $\omega-$subalgebras of $\mathbb{S}_\omega$ are connected by the zero class element $\overline{\omega}$.

4. The positive $\omega-$elements, the negative $\omega-$elements and the balanced elements are called signed and denoted by $\mathbb{S}_\omega^{\vee} = \mathbb{S}_\omega^{(+)} \cup \mathbb{S}_\omega^{(-)}$, where the zero class $\overline{(\omega,\omega)}$ corresponds to ω.

2.3 Rules of calculation in omega

Notation 11. [6] *Let $a \in \mathbb{A}_\omega$. Then we admit the following notations:*

$$+a = \overline{(a,\omega)}, \; -a = \overline{(\omega,a)}, \; \cdot a = \overline{(a,a)}. \tag{16}$$

By results in Proposition 7 and Proposition 8 and the above notation, it is easy to verify the rules of calculation in the following proposition:

Proposition 12. [6] *For all $a, b \in A_\omega$, we have*

i. $(+a) \oplus (+b) = +(a \oplus b)$;

ii. $(+a) \oplus (-b) = \begin{cases} +a & \text{if } b < a \\ -b & \text{if } b > a; \\ \cdot a & \text{if } b - a \end{cases}$

iii. $(\pm a) \oplus (\cdot b) = \begin{cases} \pm a & \text{if } b < a \\ \cdot b & \text{if } b > a \end{cases}$:

iv. $(-a) \oplus (-b) = -(a \oplus b)$;

v. $(+a) \otimes (+b) = +(a \otimes b)$;

vi. $(+a) \otimes (-b) = -(a \otimes b)$;

vii. $(\pm a) \otimes (\cdot b) = \cdot(a \otimes b)$;

viii. $(-a) \otimes (-b) = +(a \otimes b)$.

From the previous rules, we can notice that the sign of the result in the addition operation follows the greater element in A_ω. While in the multiplication operation, the balance sign is the strong one (has priority).

3. Symmetrized omega topology

In this section, we define a new topology on symmetrized omega algebra and discuss some of its topological properties.

Throughout this paper, we assume that $\otimes|A = \circ$.

Proposition 13. Let $\mathbb{S}_\omega = \left(\frac{\mathcal{P}_\omega}{\sim}, \overline{\oplus}, \overline{\otimes}, \overline{\omega}, \overline{e}\right)$ be a symmetrized ω−algebra over the abelian group $G \times G$, where $\mathcal{P}_\omega = A_\omega \times A_\omega$ and $\otimes|A = \circ$. We define a new topology on $\mathbb{S}_\omega$ called a symmetrized omega topology, denoted by τ_ω as follow:

$\tau_\omega = \{\varnothing, \mathbb{S}_\omega\} \cup \{U \subseteq \mathbb{S}_\omega : \mathbb{S}_\omega^{(0)} \subseteq U$ *and for any* $+a, -a \in U$, *their multiplicative inverses exists in U, where* $a \in A_\omega \backslash \{\omega\}\}$.

Proof. Condition $\varnothing, \mathbb{S}_\omega \in \tau_\omega$ is satisfied from the definition of τ_ω. Now let $V_1, V_2 \in \tau_\omega$ be arbitrary. If either V_1 or V_2 is equal $\varnothing$, then $V_1 \cap V_2 = \varnothing \in \tau_\omega$. Assume now, $V_1 \neq \varnothing \neq V_2$. If either V_1 or V_2 is equal $\mathbb{S}_\omega$, then $V_1 \cap V_2 = V_1 \, or \, V_2 \in \tau_\omega$. So assume that, $V_1 \neq \mathbb{S}_\omega \neq V_2$, then $V_1 \cap V_2 \in \tau_\omega$, because $\mathbb{S}_\omega^{(0)} \subseteq V_1$ and $\mathbb{S}_\omega^{(0)} \subseteq V_2$, hence $\mathbb{S}_\omega^{(0)} \subseteq V_1 \cap V_2$, also for any element $+a, -a \in V_1 \cap V_2$, where $a \neq \omega$, then we have $+a, -a \in V_1$ and $+a, -a \in V_2$, then the multiplicative inverse of $+a, -a$ must belong to V_1 and V_2. Hence, the multiplicative inverse of $+a, -a$ belong to $V_1 \cap V_2$, then $V_1 \cap V_2 \in \tau_\omega$. For the third condition let $S_\gamma \in \tau_\omega$ for any $\gamma \in \Lambda$. If $S_\gamma = \varnothing$ for all $\gamma \in \Lambda$, then $\cup_{\gamma \in \Lambda} S_\gamma = \varnothing \in \tau_\omega$. So, assume that some member is nonempty, but since the empty set does not affect any union, we may assume, without loss of generality, that $S_\gamma \neq \varnothing$ for all $\gamma \in \Lambda$. If there exist $\gamma_1 \in \Lambda$ such that $S_{\gamma_1} = \mathbb{S}_\omega$, then $\cup_{\gamma \in \Lambda} S_\gamma = \mathbb{S}_\omega \in \tau_\omega$. So, assume now that $S_\gamma \neq \mathbb{S}_\omega$ for all $\gamma \in \Lambda$. Then $\cup_{\gamma \in \Lambda} S_\gamma \in \tau_\omega$, because $\mathbb{S}_\omega^{(0)} \subseteq S_\gamma$ for all $\gamma \in \Lambda$. Hence $\mathbb{S}_\omega^{(0)} \subseteq \cup_{\gamma \in \Lambda} S_\gamma$. Also for any $+a, -a \in \cup_{\gamma \in \Lambda} S_\gamma$, where $a \neq \omega$, there exists $\gamma_1, \gamma_2 \in \Lambda$ such that $+a \in S_{\gamma_1}$ and $-a \in S_{\gamma_2}$. Hence the multiplicative inverse of $+a, -a$ belong to S_{γ_1} and S_{γ_2} respectively, then the multiplicative inverse of $+a, -a$ belong to $\cup_{\gamma \in \Lambda} S_\gamma$. Hence $\cup_{\gamma \in \Lambda} S_\gamma \in \tau_\omega$.

Therefore, $(\mathbb{S}_\omega, \tau_\omega)$ is topological space.

Proposition 14. *If* $\mathbb{S}_\omega = \left(\frac{\mathcal{P}_\omega}{\sim}, \overline{\oplus}, \overline{\otimes}, \overline{\omega}, \overline{e}\right)$ *be a symmetrized* ω*−algebra over the abelian group* $G \times G$, *where* $\mathcal{P}_\omega = A_\omega \times A_\omega$, *and* $\otimes|A = \circ$. *Then an element a has a multiplicative inverse in* A_ω *if and only if the elements* $+a, -a$ *have a multiplicative inverses in* $\mathbb{S}_\omega$.

Proof. Let $a \in A_\omega$ be arbitrary, which has a multiplicative inverse, denoted by a^{-1}, then

$$\begin{aligned}(+a) \otimes (+a^{-1}) = \overline{(a, \omega)} \otimes \overline{(a^{-1}, \omega)} &= \overline{((a \otimes a^{-1}) \oplus (\omega \otimes \omega), (a \otimes \omega) \oplus (\omega \otimes a^{-1}))} \\ &= \overline{(a \otimes a^{-1}, \omega)} \\ &= \overline{(a \circ a^{-1}, \omega)} = \overline{(e, \omega)} = \overline{e},\end{aligned}$$

then $+a^{-1}$ is a multiplicative inverse of $+a$ in $\mathbb{S}_\omega$. Also,

$$\begin{aligned}(-a) \otimes (-a^{-1}) = \overline{(\omega, a)} \otimes \overline{(\omega, a^{-1})} &= \overline{((\omega \otimes \omega) \oplus (a \otimes a^{-1}), (\omega \otimes a^{-1}) \oplus (a \otimes \omega))} \\ &= \overline{(a \otimes a^{-1}, \omega)} \\ &= \overline{(a \circ a^{-1}, \omega)} = \overline{(e, \omega)} = \overline{e},\end{aligned}$$

then $-a^{-1}$ is a multiplicative inverse of $-a$ in $\mathbb{S}_\omega$.

Conversely, let $+a \in \mathbb{S}_\omega$ be arbitrary, which has a multiplicative inverse $\overline{(x,y)}$, where $x,y \in A_\omega$, then we have:

$$\begin{aligned}(+a)\otimes\overline{(x,y)} = \overline{(a,\omega)}\otimes\overline{(x,y)} &= \overline{((a\otimes x)\oplus(\omega\otimes y),(a\otimes y)\oplus(\omega\otimes x))}\\ &= \overline{(a\otimes x, a\otimes y)}\\ &= \overline{(a\circ x, a\circ y)} = \overline{(e,\omega)} = \bar{e},\end{aligned} \tag{17}$$

then $a\circ x = e$ and $a\circ y = \omega$. Hence, $x = a^{-1}$ is the multiplicative of a in A_ω.

Let $-a \in \mathbb{S}_\omega$ be arbitrary, which has a multiplicative inverse $\overline{(x,y)}$, where $x,y \in A_\omega$, then we have:

$$\begin{aligned}(-a)\otimes\overline{(x,y)} = \overline{(\omega,a)}\otimes\overline{(x,y)} &= \overline{((\omega\otimes x)\oplus(a\otimes y),(\omega\otimes y)\oplus(a\otimes x))}\\ &= \overline{(a\otimes y, a\otimes x)}\\ &= \overline{(a\circ y, a\circ x)} = \overline{(e,\omega)} = \bar{e},\end{aligned} \tag{18}$$

then $a\circ y = e$ and $a\circ x = \omega$. Hence, $y = a^{-1}$ is the multiplicative inverse of a in A_ω.

Proposition 15. *For any $\cdot a \in \mathbb{S}_\omega^{(0)}$, where $\omega \neq e$, then $\cdot a$ has no multiplicative inverse.*

Proof. Suppose that, $\cdot a \in \mathbb{S}_\omega^{(0)}$ has a multiplicative inverse $\overline{(x,y)}$, where $x,y \in A_\omega$, then

$$\begin{aligned}(\cdot a)\otimes\overline{(x,y)} = \overline{(a,a)}\otimes\overline{(x,y)} &= \overline{((a\otimes x)\oplus(a\otimes y),(a\otimes y)\oplus(a\otimes x))}\\ &= \overline{((a\circ x)\oplus(a\circ y),(a\circ y)\oplus(a\circ x))} = \overline{(e,\omega)}.\end{aligned} \tag{19}$$

Hence, $(a\circ x)\oplus(a\circ y) = e$ and $(a\circ y)\oplus(a\circ x) = \omega$, thus a contradiction.

Corollary 16. *If $a \in A_\omega\backslash\{\omega\}$ has no multiplicative inverse, then $\mathbb{S}_\omega$ is the only open set in $(\mathbb{S}_\omega, \tau_\omega)$ containing $+a$ and $-a$..*

Remark 17.

1. We denote for any element $a \in \mathbb{S}_\omega$, by $sign(.)a$ or $sign(a)a$, where $sign(.), sign(a) \in \{+, -, \cdot\}$;
2. If $a = \omega$, then $\cdot a = +a = -a$;
3. If a^{-1} is the multiplicative inverse of a in A_ω, then $+a^{-1}$ and $-a^{-1}$ are the multiplicative inverses of $+a$ and $-a$, respectively in $\mathbb{S}_\omega$ (vice versa);
4. If a has no multiplicative inverse in A_ω, then $+a$ and $-a$ have no multiplicative inverses in $\mathbb{S}_\omega$ (vice versa).

Proposition 18. *A symmetrized omega topological space $(\mathbb{S}_\omega, \tau_\omega)$ has a base*

$$\mathfrak{B} = \Big\{\mathbb{S}_\omega, \mathbb{S}_\omega^{(0)}, \mathbb{S}_\omega^{(0)} \cup \{+a, +a^{-1}\}, \mathbb{S}_\omega^{(0)} \cup \{-a, -a^{-1}\} : a \in A_\omega\backslash\{\omega\} \text{ has a multiplicative inverse}\Big\}. \tag{20}$$

Proof. For the first condition, let $B \in \mathfrak{B}$ be arbitrary. If $B = \mathbb{S}_\omega^{(0)}$ or $\mathbb{S}_\omega$ then $B \in \tau_\omega$ (satisfied by the definition of τ_ω). Assuming that,

$B = \mathbb{S}_\omega^{(0)} \cup \{+a, +a^{-1}\}$ *or* $\mathbb{S}_\omega^{(0)} \cup \{-a, -a^{-1}\}$ for any $a \in A_\omega \backslash \{\omega\}$, which has a multiplicative inverse in A_ω, then $B \in \tau_\omega$, because $\mathbb{S}_\omega^{(0)} \subset B$, and the elements $+a$ *and* $-a$ in B its multiplicative inverse $+a^{-1}$ *and* $-a^{-1}$ respectively, exists in B. Thus $\mathfrak{B} \subseteq \tau_\omega$. For the second condition, let $sign(a)a \in \mathbb{S}_\omega$ be arbitrary. Let U be any open neighborhood of $sign(a)a$ in $\mathbb{S}_\omega$. Then we have three cases:

Case 1: If $sign(a) = \cdot$, then there exists $B = \mathbb{S}_\omega^{(0)} \in \mathfrak{B}$, such that $\cdot a \in B \subseteq U$, because the smallest open neighborhood in $\mathbb{S}_\omega$ containing $\cdot a$ is $\mathbb{S}_\omega^{(0)}$.

Case 2: If $sign(a) = +$, where $a \neq \omega$ (If $a = \omega$, then we have $+\omega = -\omega = \cdot\omega$, this is Case 1),

Subcase 2.1: If a has a multiplicative inverse in A_ω, then there exists $B = \mathbb{S}_\omega^{(0)} \cup \{+a, +a^{-1}\} \in \mathfrak{B}$, such that $+a \in B \subseteq U$, because the smallest open neighborhood in $\mathbb{S}_\omega$ containing $+a$ is $\mathbb{S}_\omega^{(0)} \cup \{+a, +a^{-1}\}$.

Subcase 2.2: If a has no multiplicative inverse in A_ω, then there exists $B = \mathbb{S}_\omega$, such that $+a \in B \subseteq U$, because the smallest open neighborhood in $\mathbb{S}_\omega$ containing $+a$ is $\mathbb{S}_\omega$.

Case 3: If $sign(a) = -$, where $a \neq \omega$.

Subcase 3.1: If a has a multiplicative inverse in A_ω, then there exists $B = \mathbb{S}_\omega^{(0)} \cup \{-a, -a^{-1}\} \in \mathfrak{B}$, such that $-a \in B \subseteq U$, because the smallest open neighborhood in $\mathbb{S}_\omega$ containing $-a$ is $\mathbb{S}_\omega^{(0)} \cup \{-a, -a^{-1}\}$.

Subcase 3.2: If a has no multiplicative inverse in A_ω, then there exists $B = \mathbb{S}_\omega$, such that $-a \in B \subseteq U$, because the smallest open neighborhood in $\mathbb{S}_\omega$ containing $-a$ is $\mathbb{S}_\omega$.

Therefore, $\mathfrak{B}$ is a base for the symmetrized omega topological space $(\mathbb{S}_\omega, \tau_\omega)$.

Corollary 19. *If* $(A_\omega \backslash \{\omega\}, \otimes)$ *be a group, then the symmetrized omega topological space* $(\mathbb{S}_\omega, \tau_\omega)$ *has a base,*

$$\mathfrak{B} = \left\{\mathbb{S}_\omega^{(0)}, \mathbb{S}_\omega^{(0)} \cup \{+a, +a^{-1}\}, \mathbb{S}_\omega^{(0)} \cup \{-a, -a^{-1}\} : a \in A_\omega \backslash \{\omega\}\right\}. \tag{21}$$

Corollary 20. *Let* $\varnothing \neq U \subseteq A_\omega$, *then* $U \in \tau_\omega$ *if and only if for each* $sign(a)a \in U$, *there exists basic open set* $B \in \mathfrak{B}$, *such that* $sign(a)a \in B \subseteq U$.

Proposition 21. *If* A_ω *has a finite number of elements, which have a multiplicative inverses, then the symmetrized omega topological space* $(\mathbb{S}_\omega, \tau_\omega)$ *is second countable.*

Proof. Suppose that $a_1, a_2, \cdots, a_m$, where $m \in \mathbb{Z}^+$ are the finite number of elements in A_ω, which have a multiplicative inverses. Then.

$\mathfrak{B} = \left\{\mathbb{S}_\omega, \mathbb{S}_\omega^{(0)}, \mathbb{S}_\omega^{(0)} \cup \{+a_1, +a_1^{-1}\}, \mathbb{S}_\omega^{(0)} \cup \{-a_1, -a_1^{-1}\}, \cdots, \mathbb{S}_\omega^{(0)} \cup \{+a_m, +a_m^{-1}\},\right.$
$\left.\mathbb{S}_\omega^{(0)} \cup \{-a_m, -a_m^{-1}\}\right\}$ is a countable base for $(\mathbb{S}_\omega, \tau_\omega)$.

Proposition 22. *The symmetrized omega topological space* $(\mathbb{S}_\omega, \tau_\omega)$ *is first countable.*

Proof. Let $sign(a)a \in \mathbb{S}_\omega$ be arbitrary. Then we have three cases:

Case 1: If $sign(a) = \cdot$, then $\mathfrak{B}(\cdot a) = \left\{\mathbb{S}_\omega^{(0)}\right\}$ is a countable local base at $\cdot a$.

Case 2: If $sign(a) = +$, where $a \neq \omega$ (If $a = \omega$, then $+\omega = -\omega = \cdot\omega$, this is Case 1),

Subcase 2.1: If a has a multiplicative inverse in A_ω, then $\mathfrak{B}(+a) = \left\{\mathbb{S}_\omega^{(0)} \cup \{+a, +a^{-1}\}\right\}$ is a countable local base at $+a$.

Subcase 2.2: If a has no multiplicative inverse in A_ω, then $\mathfrak{B}(+a) = \{\mathbb{S}_\omega\}$ is a countable local base at $+a$.

Case 3: If $sign(a) = -$, where $a \neq \omega$,

Subcase 3.1: If a has a multiplicative inverse in A_ω, then $\mathfrak{B}(-a) = \left\{\mathbb{S}_\omega^{(0)} \cup \{-a, -a^{-1}\}\right\}$ is a countable local base at $-a$.

Subcase 3.2: If a has no multiplicative inverse in A_ω, then $\mathfrak{B}(-a) = \{\mathbb{S}_\omega\}$ is a countable local base at $-a$. Hence, for any $sign(a)a \in \mathbb{S}_\omega$, there exists a countable local base at $sign(a)a$.

Therefore, $(\mathbb{S}_\omega, \tau_\omega)$ is first countable.

Proposition 23. *The symmetrized omega topological space* $(\mathbb{S}_\omega, \tau_\omega)$ *is separable.*

Proof. There exists $\{\cdot\omega\} = \left\{\overline{(\omega, \omega)}\right\} \subseteq \mathbb{S}_\omega$, such that for any $U \in \tau_\omega$, we have $U \cap \{\cdot\omega\} \neq \varnothing$, because any open set in $(\mathbb{S}_\omega, \tau_\omega)$ must be containing $\mathbb{S}_\omega^{(0)}$, and $\cdot\omega \in \mathbb{S}_\omega^{(0)}$. Then $\{\cdot\omega\}$ is countable dense subset of $\mathbb{S}_\omega$. Therefore, $(\mathbb{S}_\omega, \tau_\omega)$ is separable.

Let us recall this definition.

Definition 24. A topological space X is said to be *hyperconnected space* if every non-empty open set of X is dense in X or there exists no disjoint non-empty open sets in X.

Proposition 25. *The symmetrized omega topological space* $(\mathbb{S}_\omega, \tau_\omega)$ *is hyperconnected.*

Proof. If $\mathbb{S}_\omega$ is singleton, then it is hyperconnected. Suppose that $\mathbb{S}_\omega$, which has more than one element. Since any nonempty open set in $\mathbb{S}_\omega$ is containing $\mathbb{S}_\omega^{(0)}$, then $\mathbb{S}_\omega$ has no disjoint nonempty open sets. Hence, $(\mathbb{S}_\omega, \tau_\omega)$ is hyperconnected.

Since any hyperconnected space is connected and locally connected, then we conclude the following corollaries.

Corollary 26. *The symmetrized omega topological space* $(\mathbb{S}_\omega, \tau_\omega)$ *is connected.*

Corollary 27. *The symmetrized omega topological space* $(\mathbb{S}_\omega, \tau_\omega)$ *is locally connected.*

Proposition 28. *Let* $(A_\omega \backslash \{\omega\}, \otimes)$ *be a group, has more than one element. Then the symmetrized omega topological space* $(\mathbb{S}_\omega, \tau_\omega)$ *is not* T_0.

Proof. If $A_\omega = \{\omega = e\}$, then $\mathbb{S}_\omega = \{\cdot\omega\}$ is singleton, we are done (because some of omega algebra, has $\omega = e$). Suppose that A_ω has more than one element. Let $a \neq \omega$, then there exist $\cdot a \neq \cdot\omega$ in $\mathbb{S}_\omega$. Let U be any open set in $\mathbb{S}_\omega$, containing either $\cdot a$ or $\cdot\omega$, by the definition of τ_ω we have $\mathbb{S}_\omega^{(0)} \subseteq U$, but $\cdot\omega$, $\cdot a \in \mathbb{S}_\omega^{(0)}$. Then there is no open set containing only $\cdot\omega$ or $\cdot a$. Hence, $(\mathbb{S}_\omega, \tau_\omega)$ is not T_0.

Proposition 29. *If* $(A_\omega \backslash \{\omega\}, \otimes)$ *be a group, has more than one element, then the symmetrized omega topological space* $(\mathbb{S}_\omega, \tau_\omega)$ *is not regular.*

Proof. There exists $K = \mathbb{S}_\omega \backslash \mathbb{S}_\omega^{(0)}$ is a closed subset of $\mathbb{S}_\omega$ and there exists $a \neq \omega$, such that $\cdot a \notin K$. We cannot separate $\cdot a$, and K by any open sets (because any open sets in $\mathbb{S}_\omega$ is containing $\mathbb{S}_\omega^{(0)}$, where $\cdot a \in \mathbb{S}_\omega^{(0)}$). Therefore, $(\mathbb{S}_\omega, \tau_\omega)$ is not regular.

Proposition 30. *If* $(A_\omega \backslash \{\omega\}, \otimes)$ *be a group, has more than one element, then the symmetrized omega topological space* $(\mathbb{S}_\omega, \tau_\omega)$ *is not normal.*

Proof. If $A_\omega = \{\omega\}$, then $\mathbb{S}_\omega = \{\cdot\omega\}$ is singleton, we are done (because some of omega algebra, we have $\omega = e$). Suppose that A_ω has more than one element. Let $a \in A_\omega \backslash \{\omega\}$. Then we have two cases:

Case 1: If $a = e$, then we have $K = \{+e\}$, and $H = \{-e\}$ are two disjoint closed subsets of $\mathbb{S}_\omega$, such that we cannot separate them by any open sets (because any nonempty open sets in $\mathbb{S}_\omega$ is containing $\mathbb{S}_\omega^{(0)}$).

Case 2: If $a \neq e$, then we have $K = \{+a, +a^{-1}\}$, and $H = \{-a, -a^{-1}\}$ are two disjoint closed subsets of $\mathbb{S}_\omega$, such that we cannot separate them by any open sets (because any nonempty open sets in $\mathbb{S}_\omega$ is containing $\mathbb{S}_\omega^{(0)}$). Therefore, $(\mathbb{S}_\omega, \tau_\omega)$ is not normal.

Proposition 31. *If* $a \in A_\omega \backslash \{\omega\}$ *has no multiplicative inverse, then the symmetrized omega topological space* $(\mathbb{S}_\omega, \tau_\omega)$ *is normal.*

Proof. Suppose that, V be any non-empty closed subset of $\mathbb{S}_\omega$. Then $+a \in V$. Suppose not, $+a \notin V$, then $+a \in \mathbb{S}_\omega \backslash V$. By the definition of τ_ω, $\mathbb{S}_\omega \backslash V$ is not open, thus a contradiction. Hence, $+a$ belong to any non-empty closed subsets of $\mathbb{S}_\omega$.

DOI: http://dx.doi.org/10.5772/intechopen.92764

Let K and H be any two disjoint closed subsets of $\mathbb{S}_\omega$. Then K or H is equal $\varnothing$. If $K = \varnothing$, then there exists $U = \varnothing$ and $V = \mathbb{S}_\omega$ are two disjoint open sets in $\mathbb{S}_\omega$ containing K and H, respectively. If $H = \varnothing$, then there exists $U = \varnothing$ and $V = \mathbb{S}_\omega$ are two disjoint open sets in $\mathbb{S}_\omega$ containing H and K, respectively. Therefore, $(\mathbb{S}_\omega, \tau_\omega)$ is normal.

Proposition 32. *If $(A_\omega \backslash \{\omega\}, \otimes)$ be a group and A is uncountable infinite set, then the symmetrized omega topological space $(\mathbb{S}_\omega, \tau_\omega)$ is not compact (Lindelöf).*

Proof. There exists $\left\{\mathbb{S}_\omega^{(0)}, \mathbb{S}_\omega^{(0)} \cup \{+a, +a^{-1}\}, \mathbb{S}_\omega^{(0)} \cup \{-a, -a^{-1}\} : a \in A_\omega \backslash \{\omega\}\right\}$, which is an open cover of $\mathbb{S}_\omega$, and has no finite (countable) subcover of $\mathbb{S}_\omega$.

Proposition 33. *Let $a \in A_\omega \backslash \{\omega\}$ has no multiplicative inverse. Then the symmetrized omega topological space $(\mathbb{S}_\omega, \tau_\omega)$ is compact.*

Proof. Let $\{C_\alpha : \alpha \in \Lambda\}$ be any open cover of $\mathbb{S}_\omega$. Since $+a \in \mathbb{S}_\omega$, then for some $\beta \in \Lambda$, there exists C_β containing $+a$. But $C_\beta = \mathbb{S}_\omega$, because $\mathbb{S}_\omega$ is the only open set containing $+a$. Hence, $\{C_\beta\}$ is a finite subcover of $\{C_\alpha : \alpha \in \Lambda\}$, which cover $\mathbb{S}_\omega$. Therefore $(\mathbb{S}_\omega, \tau_\omega)$ is a compact space.

Since any compact space is Lindelöf and countably compact, then we conclude the following corollaries.

Corollary 34. *If $a \in A_\omega \backslash \{\omega\}$ has no multiplicative inverse, then the symmetrized omega topological space $(\mathbb{S}_\omega, \tau_\omega)$ is Lindelöf.*

Corollary 35. *If $a \in A_\omega \backslash \{\omega\}$ has no multiplicative inverse, then the symmetrized omega topological space $(\mathbb{S}_\omega, \tau_\omega)$ is countably compact.*

Remark 36. Since every nonempty open sets in $(\mathbb{S}_\omega, \tau_\omega)$ contains $\mathbb{S}_\omega^{(0)}$. Then the closure of any nonempty open sets is equal $\mathbb{S}_\omega$.

4. Some of the fundamental properties for different examples on symmetrized omega topology

In this section, we give two different examples of symmetrized omega topologies. The examples are from an ordered infinite set.

Example 37. By Example 4, we set $W = \{0, 1, 2, 3, \cdots\}$. Then $(\mathbb{S}_{-\infty}, \tau_{-\infty})$, which is topological space, where $\mathbb{S}_{-\infty} = \left(\frac{\mathcal{P}_{-\infty}}{\sim}, \overline{\oplus}, \overline{\otimes}, \overline{-\infty}, \overline{0}\right)$ be a symmetrized $-\infty-$algebra over the abelian group $\mathbb{Z} \times \mathbb{Z}$ and $\mathcal{P}_{-\infty} = W_{-\infty} \times W_{-\infty}$. Let $a \in W \backslash \{0\}$ be arbitrary. Then $+a^{-1}$ and $-a^{-1}$ are not exists in $\mathbb{S}_{-\infty}$, where $+a^{-1}$ and $-a^{-1}$ are the multiplicative inverses of $+a$ and $-a$ in $\mathbb{S}_{-\infty}$ respectively (because a in $W_{-\infty}$ has no multiplicative inverse). If $a = 0$, then $+0^{-1} = +0$ and $-0^{-1} = -0$ (because the multiplicative inverse of 0 in $W_{-\infty}$ is 0, that is $0^{-1} = 0$). Hence,

$$\tau_{-\infty} = \left\{\mathbb{S}_{-\infty}, \varnothing, \mathbb{S}_{-\infty}^{(0)}, \mathbb{S}_{-\infty}^{(0)} \cup \{+0\}, \mathbb{S}_{-\infty}^{(0)} \cup \{-0\}, \mathbb{S}_{-\infty}^{(0)} \cup \{+0, -0\}\right\}. \quad (22)$$

A direct check shows that $(\mathbb{S}_{-\infty}, \tau_{-\infty})$ is a topological space.

Proposition 38. *The symmetrized omega topological space $(\mathbb{S}_{-\infty}, \tau_{-\infty})$ is Second countable.*

Proof. There exists only one element $0 \in W_{-\infty}$, which has a multiplicative inverse, then by Proposition 21, $(\mathbb{S}_{-\infty}, \tau_{-\infty})$ is second countable.

Since any second countable space is first countable and separable, then we conclude the following corollaries.

Corollary 39. *The symmetrized omega topological space $(\mathbb{S}_{-\infty}, \tau_{-\infty})$ is first countable.*

Corollary 40. *The symmetrized omega topological space $(\mathbb{S}_{-\infty}, \tau_{-\infty})$ is separable.*

Proposition 41. *The symmetrized omega topological space $(\mathbb{S}_{-\infty}, \tau_{-\infty})$ is not T_0.*

Proof. There exists $+2 \neq +3$ in $\mathbb{S}_{-\infty}$. Let U be any open set, which either containing $+2$ or $+3$. However, there exists only one open set $U = \mathbb{S}_{-\infty}$ containing $+2, +3$. Hence, $(\mathbb{S}_{-\infty}, \tau_{-\infty})$ is not T_0.

Proposition 42. *The symmetrized omega topological space* $(\mathbb{S}_{-\infty}, \tau_{-\infty})$ *is not regular.*

Proof. There exists a closed set $K = \mathbb{S}_{-\infty} \backslash \left(\mathbb{S}^{(0)}_{-\infty} \cup \{+0\} \right)$ and $+0 \notin K$, such that $+0$ and K cannot separate by any two disjoint open sets. Hence, $(\mathbb{S}_{-\infty}, \tau_{-\infty})$ is not regular.

Proposition 43. *The symmetrized omega topological space* $(\mathbb{S}_{-\infty}, \tau_{-\infty})$ *is normal.*

Proof. There exists an element $2 \in W_{-\infty} \backslash \{-\infty\}$, which has no multiplicative inverse, then by Proposition 31, $(\mathbb{S}_{-\infty}, \tau_{-\infty})$ is a normal space.

Proposition 44. *The symmetrized omega topological space* $(\mathbb{S}_{-\infty}, \tau_{-\infty})$ *is hyperconnected.*

Proof. Since any nonempty open set in $\mathbb{S}_{-\infty}$ is containing $\mathbb{S}^{(0)}_{-\infty}$, then $\mathbb{S}_{-\infty}$ has no disjoint nonempty open sets. Hence, $(\mathbb{S}_{-\infty}, \tau_{-\infty})$ is hyperconnected.

Since any hyperconnected space is connected and locally connected, then we conclude the following corollaries.

Corollary 45. *The symmetrized omega topological space* $(\mathbb{S}_{-\infty}, \tau_{-\infty})$ *is connected.*

Corollary 46. *The symmetrized omega topological space* $(\mathbb{S}_{-\infty}, \tau_{-\infty})$ *is locally connected.*

Proposition 47. *The symmetrized omega topological space* $(\mathbb{S}_{-\infty}, \tau_{-\infty})$ *is compact.*

Proof. There exists an element $2 \in W_{-\infty} \backslash \{-\infty\}$, which has no multiplicative inverse. Hence by Proposition 33, $(\mathbb{S}_{-\infty}, \tau_{-\infty})$ is compact.

Since any compact space is Lindelöf and countably compact, then we conclude the following corollaries.

Corollary 48. *The symmetrized omega topological space* $(\mathbb{S}_{-\infty}, \tau_{-\infty})$ *is countably compact.*

Corollary 49. *The symmetrized omega topological space* $(\mathbb{S}_{-\infty}, \tau_{-\infty})$ *is Lindelöf.*

Example 50. In the ring $(\mathbb{R}, +, \cdot)$, we have $(\mathbb{R}, +)$ is an additive submonoid of an abelian group $(\mathbb{R}, +)$. Let $\omega = -\infty$, $a \oplus b = \max(a, b)$ and $a \otimes b = a + b, \forall a, b \in \mathbb{R}$. Then $\mathbb{R}_{-\infty} = (\mathbb{R}_{-\infty}, \oplus, \otimes, -\infty, 0)$ is $-\infty-$ algebra over the ring $(\mathbb{R}, +, \cdot)$. We have $\mathbb{S}_{-\infty} = \left(\frac{\mathcal{P}_{-\infty}}{\sim}, \overline{\oplus}, \overline{\otimes}, \overline{-\infty}, \overline{0}\right)$ be a symmetrized $-\infty-$algebra over the abelian group $\mathbb{R} \times \mathbb{R}$ and $\mathcal{P}_{-\infty} = \mathbb{R}_{-\infty} \times \mathbb{R}_{-\infty}$. Then, using the same proof as that Proposition 13. Therefore, $(\mathbb{S}_{-\infty}, \tau_{-\infty})$ is a topological space.

Remark 51. The symmetrized omega topological space $(\mathbb{S}_{-\infty}, \tau_{-\infty})$ is first countable, separable, hyperconnected, connected and locally connected and does not satisfy any of these T_0, regular, normal, Lindelöf and compact.

Example 52. In the ring $(\mathbb{R}, +, \cdot)$, we have $(\mathbb{R}, +)$ is an additive submonoid of an abelian group $(\mathbb{R}, +)$. Let $\omega = +\infty$, $a \oplus b = \min(a, b)$ and $a \otimes b = a + b, \forall a, b \in \mathbb{R}$. Then, $\mathbb{R}_{+\infty} = (\mathbb{R}_{+\infty}, \oplus, \otimes, +\infty, 0)$ is $+\infty-$ algebra over the ring $(\mathbb{R}, +, \cdot)$. We have $\mathbb{S}_{+\infty} = \left(\frac{\mathcal{P}_{+\infty}}{\sim}, \overline{\oplus}, \overline{\otimes}, \overline{+\infty}, \overline{0}\right)$ be a symmetrized $+\infty-$algebra over the abelian group $\mathbb{R} \times \mathbb{R}$ and $\mathcal{P}_{+\infty} = \mathbb{R}_{+\infty} \times \mathbb{R}_{+\infty}$. Then, using the same proof as that Proposition 13. Therefore, $(\mathbb{S}_{+\infty}, \tau_{+\infty})$ is a topological space.

Proposition 53. *The symmetrized omega topological spaces* $(\mathbb{S}_{-\infty}, \tau_{-\infty})$ *and* $(\mathbb{S}_{+\infty}, \tau_{+\infty})$ *are homeomorphic.*

Proof. There exists a map $h : (\mathbb{S}_{-\infty}, \tau_{-\infty}) \to (\mathbb{S}_{+\infty}, \tau_{+\infty})$ is defined by:

$$h(sign(a)a) = \begin{cases} sign(a)a & if \quad a \in \mathbb{R} \\ sign(-\infty)(+\infty) & if \quad sign(a)a = sign(-\infty)(-\infty) \end{cases}; \tag{23}$$

Let $sign(a)a, sign(b)b \in \mathbb{S}_{-\infty}$ be arbitrary. Let $h(sign(a)a) = h(sign(b)b)$, then $sign(a)a = sign(b)b$. Hence, h is an injective. Let $sign(a)a \in \mathbb{S}_{+\infty}$ is arbitrary, then there exists a $sign(a)a \in \mathbb{S}_{-\infty}$, such that $h(sign(a)a) = sign(a)a$. Hence, h is surjective.

Let $B \in \tau_{+\infty}$ be any basic open set. By Proposition 18, we have $\mathfrak{B} = \left\{\mathbb{S}^{(0)}_{-\infty}, \mathbb{S}^{(0)}_{-\infty} \cup \{+a, +a^{-1}\}, \mathbb{S}^{(0)}_{-\infty} \cup \{-a, -a^{-1}\} : a \in \mathbb{R}\right\}$ and $\mathfrak{B} = \left\{\mathbb{S}^{(0)}_{+\infty}, \mathbb{S}^{(0)}_{+\infty} \cup \{+a, +a^{-1}\}, \mathbb{S}^{(0)}_{+\infty} \cup \{-a, -a^{-1}\} : a \in \mathbb{R}\right\}$ are a base for $\mathbb{R}_{-\infty}$ and $\mathbb{R}_{+\infty}$, respectively.

To prove that h is continuous, we have three cases:

Case 1: If $B = \mathbb{S}^{(0)}_{+\infty}$, then $h^{-1}(B) = h^{-1}\left(\mathbb{S}^{(0)}_{+\infty}\right) = \mathbb{S}^{(0)}_{-\infty} \in \tau_{-\infty}$.

Case 2: If $B = \mathbb{S}^{(0)}_{+\infty} \cup \{+a, +a^{-1}\}$, then $h^{-1}(B) = h^{-1}\left(\mathbb{S}^{(0)}_{+\infty} \cup \{+a, +a^{-1}\}\right) = \mathbb{S}^{(0)}_{-\infty} \cup \{+a, +a^{-1}\} \in \tau_{-\infty}$.

Case 3: If $B = \mathbb{S}^{(0)}_{+\infty} \cup \{-a, -a^{-1}\}$, then $h^{-1}(B) = h^{-1}\left(\mathbb{S}^{(0)}_{+\infty} \cup \{-a, -a^{-1}\}\right) = \mathbb{S}^{(0)}_{-\infty} \cup \{-a, -a^{-1}\} \in \tau_{-\infty}$. Hence, h is continuous.

To prove that h^{-1} is continuous, we have three cases: (since h is one to one and onto, then $\left(h^{-1}\right)^{-1}(B) = h(B)$).

Case 1: If $B = \mathbb{S}^{(0)}_{-\infty}$, then $\left(h^{-1}\right)^{-1}(B) = h(B) = h\left(\mathbb{S}^{(0)}_{-\infty}\right) = \mathbb{S}^{(0)}_{+\infty} \in \tau_{+\infty}$.

Case 2: If $B = \mathbb{S}^{(0)}_{-\infty} \cup \{+a, +a^{-1}\}$, then $\left(h^{-1}\right)^{-1}(B) = h(B) = h\left(\mathbb{S}^{(0)}_{-\infty} \cup \{+a, +a^{-1}\}\right) = \mathbb{S}^{(0)}_{+\infty} \cup \{+a, +a^{-1}\} \in \tau_{+\infty}$.

Case 3: If $B = \mathbb{S}^{(0)}_{-\infty} \cup \{-a, -a^{-1}\}$, then $\left(h^{-1}\right)^{-1}(B) = h(B) = h\left(\mathbb{S}^{(0)}_{-\infty} \cup \{-a, -a^{-1}\}\right) = \mathbb{S}^{(0)}_{+\infty} \cup \{-a, -a^{-1}\} \in \tau_{+\infty}$. Hence h^{-1} is continuous (which means h is open).

Therefore, h is homeomorphism, then $(\mathbb{S}_{-\infty}, \tau_{-\infty})$ and $(\mathbb{S}_{+\infty}, \tau_{+\infty})$ are homeomorphic.

5. Symmetrized omega topology and other properties

Recall that a subset A of a space X is said to be *regularly-open* or an *open domain* if it is the interior of its own closure (see [7]). A set A is said to be a *regularly-closed* or a *closed domain* if its complement is an open domain. A subset A of a space X is called a *π-closed* if it is a finite intersection of closed domain sets (see [8]). A subset A is called a *π-open* if its complement is a π-closed. If $\mathcal{T}$ and $\mathcal{T}'$ are two topologies on a set X such that $\mathcal{T}' \subseteq \mathcal{T}$, then $\mathcal{T}'$ is called the coarser topology than $\mathcal{T}$, and $\mathcal{T}$ is called the finer. A space X is *π-normal* [9] if any pair of disjoint closed subsets A and B of X, one of which is π-closed, can be separated by two disjoint open subsets. A space X is *almost-normal* [9] if any pair of disjoint closed subsets A and B of X, one of which is a closed domain, can be separated by two disjoint open subsets. A space X is *mildly normal* [10] if any pair of disjoint closed domain subsets A and B of X can be separated by two disjoint open subsets. A space $(X, \mathcal{T})$ is *epi-mildly normal* [11] if there exists a coarser topology $\mathcal{T}'$ on X such that $(X, \mathcal{T}')$ is T_2 and mildly normal space. A space $(X, \mathcal{T})$ is *epi-almost normal* [12] if there exists a coarser topology $\mathcal{T}'$ on X such that $(X, \mathcal{T}')$ is T_2 and almost normal space.

Theorem 54. *If $(A_\omega \backslash \{\omega\}, \otimes)$ be a group has more than one element, then the symmetrized omega topological space $(\mathbb{S}_\omega, \tau_\omega)$ is π-normal.*

Proof. Since the only π-closed sets are the ground set S_ω and the empty set, then (S_ω, τ_ω) is a π-normal.

It is clear from the definitions that

$$normal \Rightarrow \pi - normal \Rightarrow almost\ normal \Rightarrow mildly\ normal. \tag{24}$$

By (24) and Theorem 54, we conclude the following Corollaries.

Corollary 55. *If* $(A_\omega\backslash\{\omega\}, \otimes)$ *be a group has more than one element, then the symmetrized omega topological space* $(\mathbb{S}_\omega, \tau_\omega)$ *is almost normal.*

Corollary 56. *If* $(A_\omega\backslash\{\omega\}, \otimes)$ *be a group has more than one element, then the symmetrized omega topological space* $(\mathbb{S}_\omega, \tau_\omega)$ *is mildly normal.*

If $(A_\omega\backslash\{\omega\}, \otimes)$ be a group has more than one element, then (S_ω, τ_ω) is not T_0 (see Proposition 28), we have the following Propositions:

Proposition 57. *If* $(A_\omega\backslash\{\omega\}, \otimes)$ *be a group has more than one element, then the symmetrized omega topological space* $(\mathbb{S}_\omega, \tau_\omega)$ *is not Epi-mildly Normal.*

Proof. Suppose that, $(\mathbb{S}_\omega, \tau_\omega)$ is Epi-mildly Normal. Then there exists a coarser topology T' on $\mathbb{S}_\omega$ such that $(\mathbb{S}_\omega, T')$ is T_2 and mildly normal space. Hence $(\mathbb{S}_\omega, \tau_\omega)$ is T_2, thus a contradiction. Then $(\mathbb{S}_\omega, \tau_\omega)$ is not Epi-mildly Normal.

Proposition 58. *If* $(A_\omega\backslash\{\omega\}, \otimes)$ *be a group has more than one element, then the symmetrized omega topological space* $(\mathbb{S}_\omega, \tau_\omega)$ *is not Epi-almost Normal.*

Proof. Using the same proof of Proposition 57.

Definition 59. Let X be a space. Then:

1. A space X is called a *C-normal* if there exist a normal space Y and a bijective function $f : X \to Y$ such that the restriction function $f|_A : A \to f(A)$ is a homeomorphism for each compact subspace $A \subseteq X$, [13].

2. A space X is called a CC-normal if there exists a normal space Y and a bijective function $f : X \to Y$ such that the restriction function $f|_A : A \to f(A)$ is a homeomorphism for each countably compact subspace $A \subseteq X$. [14].

3. A space X is called an L-normal if there exist a normal space Y and a bijective function $f : X \to Y$ such that the restriction function $f|_A : A \to f(A)$ is a homeomorphism for each lindelöf subspace $A \subseteq X$, [15].

4. A space X is called an S- normal if there exist a normal space Y and a bijective function $f : X \to Y$ such that the restriction function $f|_A : A \to f(A)$ is a homeomorphism for each separable subspace $A \subseteq X$, [16].

5. A space X is called a C-paracompact if there exist a paracompact space Y and a bijective function $f : X \to Y$ such that the restriction function $f|_A : A \to f(A)$ is a homeomorphism for each compact subspace $A \subseteq X$, [17].

6. A space X is called a C_2-paracompact if there exist a Hausdorff paracompact space Y and a bijective function $f : X \to Y$ such that the restriction function $f|_A : A \to f(A)$ is a homeomorphism for each compact subspace $A \subseteq X$, [17].

Proposition 60. *If* $a \in A_\omega\backslash\{\omega\}$ *has no multiplicative inverse, then the symmetrized omega topological space* $(\mathbb{S}_\omega, \tau_\omega)$ *is C-normal.*

Proof. By Proposition 31, $\mathbb{S}_\omega$ is a normal space. Then there exist $Y = \mathbb{S}_\omega$ is a normal space and the identity function $id : \mathbb{S}_\omega \to \mathbb{S}_\omega$ is bijective. Let C be any compact subset of $(\mathbb{S}_\omega, \tau_\omega)$. Then the restriction function $id\restriction_C : C \to f(C)$ is a homeomorphism. Therefore, $(\mathbb{S}_\omega, \tau_\omega)$ is a $C-$normal.

Since any normal space is CC-normal, L-normal and S-normal, just by taking $X = Y$ and f to be the identity function. Hence, we conclude the following Propositions.

Proposition 61. *If* $a \in A_\omega\backslash\{\omega\}$ *has no multiplicative inverse, then the symmetrized omega topological space* $(\mathbb{S}_\omega, \tau_\omega)$ *is CC-normal.*

Proof. Using the same proof of Proposition 60.

Proposition 62. *If $a \in A_\omega \backslash \{\omega\}$ has no multiplicative inverse, then the symmetrized omega topological space $(\mathbb{S}_\omega, \tau_\omega)$ is L-normal.*

Proof. Using the same proof of Proposition 60.

Proposition 63. *If $a \in A_\omega \backslash \{\omega\}$ has no multiplicative inverse, then the symmetrized omega topological space $(\mathbb{S}_\omega, \tau_\omega)$ is S-normal.*

Proof. Using the same proof of Proposition 60.

Example 64. By Example 37, $(\mathbb{S}_{-\infty}, \tau_{-\infty})$ is C-normal, CC-normal, L-normal and S-normal.

Theorem 65. *If $(A_\omega \backslash \{\omega\}, \otimes)$ be a group has more than one element, then the symmetrized omega topological space $(\mathbb{S}_\omega, \tau_\omega)$ is not S-normal.*

Proof. From the proposition any separable S-normal must be normal (see [16]) and since $(\mathbb{S}_\omega, \tau_\omega)$ is separable and not normal (see Propositions 30, 23, respectively), then $(\mathbb{S}_\omega, \tau_\omega)$ is not S-normal.

Example 66. By Example 50, $(\mathbb{S}_{-\infty}, \tau_{-\infty})$ is not a S-normal.

Theorem 67. *The symmetrized omega topological space $(\mathbb{S}_\omega, \tau_\omega)$ is not C_2-paracompact.*

Proof. Since any C_2-paracompact Fre′chet space is Hausdorff (see [17]) and $(\mathbb{S}_\omega, \tau_\omega)$ is First countable and not a Hausdorff space, $(\mathbb{S}_\omega, \tau_\omega)$ cannot be C_2-paracompact.

Theorem 68. *Let $a \in A_\omega \backslash \{\omega\}$ has no multiplicative inverse. Then the symmetrized omega topological space $(\mathbb{S}_\omega, \tau_\omega)$ is not C-paracompact.*

Proof. Assume that $(\mathbb{S}_\omega, \tau_\omega)$ is C-paracompact. Let Y be a paracompact space and $f : \mathbb{S}_\omega \to Y$ be bijective such that the restriction $f \restriction_C : C \to f(C)$ is a homeomorphism for all compact subspace C of $(\mathbb{S}_\omega, \tau_\omega)$. Hence, $\mathbb{S}_\omega \equiv Y$, since $\mathbb{S}_\omega$ is compact (see Proposition 33). However, $\mathbb{S}_\omega$ is paracompact, thus a contradiction. Because any paracompact space is Hausdorff space and $\mathbb{S}_\omega$ is not a Hausdorff space. Therefore, $(\mathbb{S}_\omega, \tau_\omega)$ is not a C-paracompact.

Author details

Mesfer Alqahtani[1], Cenap Özel[1*] and Ibtesam Alshammari[2]

1 Department of Mathematics, Faculty of Science, King Abdulaziz University, Jeddah, Saudi Arabia

2 Department of Mathematics, Faculty of Science, University of Hafr Al Batin, Hafr Al Batin, Saudi Arabia

*Address all correspondence to: cozel@kau.edu.sa

References

[1] Litvinov GL. The Maslov dequantization, idempotent and tropical mathematics: A brief introduction. Journal of Mathematical Sciences. 2007; **140**(3):426-444

[2] Maclagan D, Sturmfels B. Introduction to Tropical Geometry, Graduate Studies in Mathematics. Vol. 161. Providence, RI: American Mathematical Society; 2015

[3] Pin J-E. Tropical semirings. In: Idempotency (Bristol, 1994). Vol. 11. Cambridge: Publ. Newton Inst., Cambridge Univ. Press; 1998. pp. 50-69

[4] Simon I. Recognizable sets with multiplicities in the tropical semiring. In: Mathematical Foundations of Computer Science (Carlsbad, 1988). Lecture Notes in Computer Science. Vol. 324. Berlin: Springer; 1988. pp. 107-120

[5] Świerczkowski S. On cyclically ordered groups. Fundamenta Mathematicae. 1959;**47**:161-166 [Accessed: 2 May 2011]

[6] Khalid Nauman S, Ozel C, Zekraoui H. Abstract omega algebra that subsumes min and max plus algebras. Turkish Journal of Mathematics and Computer Science. 2019;**11**(Special Issue):1-10

[7] Kuratowski C. Topology I. 14th ed. French, New York: Hafner; 1958

[8] Zaitsev V. On certain classes of topological spaces and their bicompactifications. Doklady Akademii Nauk SSSR. 1968;**178**:778-779

[9] Kalantan L. π-normal topological spaces. Univerzitet u Nišu. 2008;**22**(1): 173-181

[10] Singal MK, Singal AR. Mildly normal spaces. Kyungpook Mathematical Journal. 1973;**13**:27-31

[11] Kalantan L, Alshammari I. Epi-mild normality. Open Mathematics. 2018;**16**: 1170-1175

[12] Alshammari I. Epi-almost normality. Journal of Mathematical Analysis. 2020;**11**(2):502-507

[13] AlZahrani S, Kalantan L. C-Normal topological property. Univerzitet u Nišu. 2017;**31**(2):407-411

[14] Kalantan L, Alhomieyed M. CC-normal topological spaces. Turkish Journal of Mathematics. 2017;(41): 749-755

[15] Kalantan L, Saeed M. L-Normality. Topology Proceedings. 2017;**50**:141-149

[16] Alhomieyed M, Kalantan L. S-Normality. Journal of Mathematical Analysis. 2018;**9**(5):48-54

[17] Saeed M, Kalantan L, Alzumi H. C-Paracompactness and C_2-paracompactness. Turkish Journal of Mathematics. 2019;**43**:9-20

Chapter 4

Some Applications of Clifford Algebra in Geometry

Ying-Qiu Gu

Abstract

In this chapter, we provide some enlightening examples of the application of Clifford algebra in geometry, which show the concise representation, simple calculation, and profound insight of this algebra. The definition of Clifford algebra implies geometric concepts such as vector, length, angle, area, and volume and unifies the calculus of scalar, spinor, vector, and tensor, so that it is able to naturally describe all variables and calculus in geometry and physics. Clifford algebra unifies and generalizes real number, complex, quaternion, and vector algebra and converts complicated relations and operations into intuitive matrix algebra independent of coordinate systems. By localizing the basis or frame of space-time and introducing differential and connection operators, Clifford algebra also contains Riemann geometry. Clifford algebra provides a unified, standard, elegant, and open language and tools for numerous complicated mathematical and physical theories. Clifford algebra calculus is an arithmetic-like operation that can be well understood by everyone. This feature is very useful for teaching purposes, and popularizing Clifford algebra in high schools and universities will greatly improve the efficiency of students to learn fundamental knowledge of mathematics and physics. So, Clifford algebra can be expected to complete a new big synthesis of scientific knowledge.

Keywords: Clifford algebra, geometric algebra, gamma matrix, multi-inner product, connection operator, Keller connection, spin group, cross ratio, conformal geometric algebra

1. A brief historical review

It is well known that a rotational transformation in the complex plane is equivalent to multiplying the complex number by a factor $e^{\theta i}$. How to generalize this simple and elegant operation to three-dimensional space is a difficult problem for many outstanding mathematicians in the early nineteenth century. William Rowan Hamilton (1805–1865) spent much of his later years studying the issue and eventually invented quaternion [1]. This generalization requires four elements $\{1, i, j, k\}$, and the spatial basis should satisfy the multiplying rules $i^2 = j^2 = k^2 = -1$ and $jk = -kj = i$, $ki = -ik = j$, and $ij = -ji = -k$. Although a quaternion is still a vector, it constitutes an associative algebra according to the above rules. However, the commutativity of multiplication is violated. Quaternion can solve the rotational transformation in three-dimensional space very well and simplify the representation of Maxwell equation system of electromagnetic field.

IntechOpen

When Hamilton introduced his quaternion algebra, German high school teacher Hermann Gunther Grassmann (1809–1877) was constructing his exterior algebra [2]. He defined the exterior product or outer product $a \wedge b$ of two vectors a and b, which satisfies anti-commutative law $a \wedge b = -b \wedge a$ and associativity $(a \wedge b) \wedge c = a \wedge (b \wedge c)$. The exterior product is a generalization of cross product in three-dimensional Euclidian space. Its geometrical meaning is the oriented volume of a parallel polyhedron. Exterior product is now the basic tool of modern differential geometry, but Grassmann's work was largely neglected in his lifetime.

British mathematician William Kingdon Clifford (1845–1879) was one of the few mathematicians who read and understood Grassmann's work. In 1878, he combined the algebraic rules of Hamilton and Grassmann to define a new algebraic system, which he himself called geometric algebra [3]. In this algebra, both the inner and exterior products of vectors can be uniquely represented by a linear combination of geometric product. In addition, geometric algebra is always isomorphic to some special matrix algebra.

Clifford algebra combines all the advantages of quaternion with the advantages of vector algebra and uniformly and succinctly describes the contents of geometry and physics. However, the vector calculus introduced by Gibbs had also successfully described the mathematical physics problem in three-dimensional space [4]. Clifford died prematurely at the age of 34, so that the theory of geometric algebra was not deeply researched and fully developed, and people still could not see the superiority of this algebra at that time. Thus, the important insights of Grassmann and Clifford were lost in the late nineteenth century papers. Mathematicians abstracted Clifford algebra from its geometric origins, and, for the most part of a century, it languished as a minor subdiscipline of mathematics and became one more algebra among so many others.

With the establishment of relativity, especially the introduction of Pauli and Dirac's matrix algebra for spin and the successful application in quantum theory [5], it was felt that there is an urgent need for a mathematical system to deal with problems in high-dimensional space-time. In the 1920s, Clifford algebra re-entered the field of vision and was paid attention and researched by some of the famous mathematicians and physicists such as R. Lipschitz, T. Vahlen, E. Cartan, E. Witt, C. Chevalley, and M. Riesz [6–8]. When only formal algebra is involved, we usually use the term "Clifford algebra," but more often use the "geometric algebra" named by Clifford himself if applied to geometric problems.

The first person who realized that Clifford algebra is a unified language in geometry and physics should be David Hestenes. By the 1960s, Hestenes began to restore the geometric meaning behind Pauli and Dirac algebra. Although his initial motivation was to gain insight into the nature of quantum mechanics, he quickly realized that Clifford algebra was a unified language and tool for mathematics, physics, and engineering. He published "space-time algebra" in 1966 and has been working on the promotion of Clifford algebra in teaching and research [9–12]. Because representation and algorithm in geometric algebra are seemingly as ordinary as arithmetic, his work has been neglected by the scientific community for more than 20 years. Only with the joint impetus of computer-aided design, computer vision and robotics, protein folding, neural networks, modern differential geometry, mathematical physics [13–17], and especially the Journal "Advances in Applied Clifford algebras" founded by Professor Jaime Keller, geometric algebra began to move towards popularity and prosperity.

As a unified and universal language of natural science, Clifford algebra is developed by many mathematicians, physicists, and engineers according to their different requirements and knowledge background. Such situation leads to "There are a thousand Hamlets in a thousand people's eyes." In this chapter, by introducing typical application of Clifford algebra in geometry, we show some special feature and elegance of the algebra.

2. Application of Clifford algebra in differential geometry

In Euclidean space, we have several important concepts such as vector, length, angle, area, volume, and tensor. The study of relationship between these concepts constitutes the whole content of Euclidean geometry. The mathematical tools previously used to discuss these contents are vector algebra and geometrical method, which are complex and require much fundamental knowledge. Clifford algebra exactly and faithfully describes the intrinsic properties of vector space by introducing concepts such as inner, exterior, and geometric products of vectors and thus becomes a unified language and standard tool for dealing with geometric and physical problems. Clifford algebra has the characteristics of simple concept, standard operation, completeness in conclusion, and easy understanding.

Definition 1 *For Minkowski space $\mathbb{M}^n$ over number field $\mathbb{F}$, if the multiplication rule of vectors satisfies*

$$1.\ \text{Antisymmetry},\quad \boldsymbol{x}\wedge\boldsymbol{y} = -\boldsymbol{y}\wedge\boldsymbol{x}; \tag{1}$$

$$2.\ \text{Associativity},\quad (\boldsymbol{x}\wedge\boldsymbol{y})\wedge\boldsymbol{z} = \boldsymbol{x}\wedge(\boldsymbol{y}\wedge\boldsymbol{z}); \tag{2}$$

$$3.\ \text{Distributivity},\quad \boldsymbol{x}\wedge(a\boldsymbol{y}+b\boldsymbol{z}) = a\boldsymbol{x}\wedge\boldsymbol{y}+b\boldsymbol{x}\wedge\boldsymbol{z},\quad a,b\in\mathbb{F}, \tag{3}$$

the algebra is called **Grassmann algebra** *and* $\boldsymbol{x}\wedge\boldsymbol{y}$ **exterior product**.

The Grassmann is also called **exterior algebra**. The geometrical meaning of $\mathbf{x}\wedge\mathbf{y}$ is oriented area of a parallelogram constructed by $\mathbf{x}$ and $\mathbf{y}$, and the geometrical meaning of $\mathbf{x}\wedge\mathbf{y}\wedge\cdots\wedge\mathbf{z}$ is the oriented volume of the parallelohedron constructed by the vectors (see **Figure 1**). We call $\mathbf{x}\wedge\mathbf{y}$ two-vector, $\mathbf{x}\wedge\mathbf{y}\wedge\mathbf{z}$ three-vector, and so on. For k-vector $\mathbf{x}\in\Lambda^k$ and l-vector $\mathbf{y}\in\Lambda^l$, we have

$$\mathbf{x}\wedge\mathbf{y} = (-1)^{kl}\mathbf{y}\wedge\mathbf{x}\in\Lambda^{k+l}.$$

By the definition, we can easily check:

Theorem 1 *For exterior algebra defined in $V = \mathbb{M}^n$, we have*

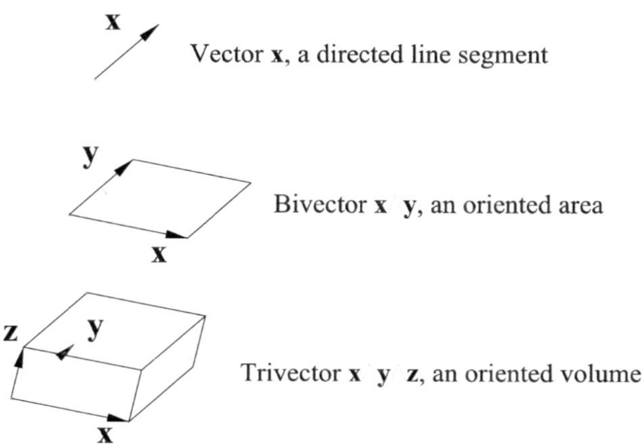

Figure 1.
Geometric meaning of exterior products of vectors.

$$\mathbb{W}^n = \mathbb{F} \oplus V \oplus \Lambda^2(V) \cdots \oplus \Lambda^n(V) = \bigoplus_{r=0}^{n} \Lambda^r(V).$$

The dimension of the algebra is

$$\dim(\mathbb{W}^n) = \sum_{k=0}^{n} C_n^k = 2^n.$$

Under the orthonormal basis $\{e_1, e_2, \cdots, e_n\}$, the exterior algebra takes the following form:

$$w = w^0 + w^k e_k + \sum_{k<l} w^{kl} e_{kl} + \sum_{j<k<l} w^{jkl} e_{jkl} + \cdots + w^{12\cdots n} e_{12\cdots n}, \tag{4}$$

in which $\forall w^{jk\cdots l} \in \mathbb{F}$, $e_{jk\cdots l} = e_j \wedge e_k \wedge \cdots \wedge e_l$, and $\forall |e_{jk\cdots l}| = 1$.

The exterior product of vectors contains alternating combinations of basis, for example:

$$\begin{aligned} V_n &= \mathbf{x}_1 \wedge \mathbf{x}_2 \wedge \cdots \wedge \mathbf{x}_n = x_1^j x_2^k \cdots x_n^l \mathbf{e}_{jk\cdots l} \\ &= \epsilon_{jk\cdots l} x_1^j x_2^k \cdots x_n^l \mathbf{e}_{12\cdots n} = \det\left(x_j^k\right) \mathbf{e}_{12\cdots n}. \end{aligned} \tag{5}$$

Definition 2 *For any vectors $x, y, z \in \mathbb{M}^n$,* **Clifford product** *of vectors is denoted by*

$$xy = x \cdot y + x \wedge y, \tag{6}$$

$$(x \wedge y)z = (y \cdot z)x - (x \cdot z)y + x \wedge y \wedge z = -(y \wedge x)z, \tag{7}$$

$$z(x \wedge y) = (x \cdot z)y - (y \cdot z)x + x \wedge y \wedge z = -z(y \wedge x), \tag{8}$$

$$(xy)z = (y \cdot z)x - (x \cdot z)y + (x \cdot y)z + x \wedge y \wedge z = x(yz). \tag{9}$$

Clifford product is also called ***geometric product****.*

Similarly, we can define Clifford algebra for many vectors as $\mathbf{xy}\cdots\mathbf{z}$. In (6), $\mathbf{x} \cdot \mathbf{y} = \eta_{ab} x^a y^b$ is the **scalar product** or **inner product** in $\mathbb{M}^n$. By $\mathbf{x} \wedge \mathbf{y} = -\mathbf{y} \wedge \mathbf{x}$, we find Clifford product is not commutative. By (6), we have

$$\mathbf{x} \cdot \mathbf{y} = \frac{1}{2}(\mathbf{xy} + \mathbf{yx}), \qquad \mathbf{x} \wedge \mathbf{y} = \frac{1}{2}(\mathbf{xy} - \mathbf{yx}), \qquad \mathbf{x} \cdot \mathbf{x} = \mathbf{xx} = \mathbf{x}^2. \tag{10}$$

Definition 3 *For Minkowski space $\mathbb{M}^{p,q}$ with metric $\eta_{ab} = diag\left(I_p, -I_q\right)$, if the Clifford product of vectors satisfies*

$$e_k e_l + e_l e_k = 2\eta_{kl}, \quad \text{or} \quad x^2 = \eta_{kl} x^k x^l,$$

then the algebra

$$c = c^0 + c^k e_k + \sum_{k<l} c^{kl} e_k e_l + \sum_{j<k<l} c^{jkl} e_j e_k e_l + \cdots + c^{12\cdots n} e_1 e_2 \cdots e_n, \tag{11}$$

is called as **Clifford algebra** *or* **geometric algebra**, *which is denoted as $C\ell_{p,q}$.*

There are several definitions for Clifford algebra [18, 19]. The above definition is the original definition of Clifford. Clifford algebra has also 2^n dimensions.

Comparing (11) with (4), we find the two algebras are isomorphic in sense of linear algebra, but their definitions of multiplication rules are different. The Grassmann products have clear geometrical meaning, but the Clifford product is isomorphic to matrix algebra and the multiplication of physical variables is Clifford product. Therefore, representing geometrical and physical variables in the form of (4) will bring great convenience [20, 21]. In this case, the relations among three products such as (6)–(9) are important.

In physics, we often use curvilinear coordinate system or consider problems in curved space-time. In this case, we must discuss problems in n dimensional pseudo Riemann manifold. At each point x in the manifold, the tangent space $T\mathbb{M}(\mathbf{x})$ is a n dimensional Minkowski space-time. The Clifford algebra can be also defined on the tangent space and then smoothly generalized on the whole manifold as follows.

Definition 4 *In $n = p + q$ dimensional manifold $T\mathbb{M}^{p,q}$ over $\mathbb{R}$, the* **element** *is defined by*

$$dx = \gamma_\mu dx^\mu = \gamma^\mu dx_\mu = \gamma_a \delta X^a = \gamma^a \delta X_a, \tag{12}$$

where γ_a is the local orthogonal frame and γ^a the coframe. The **distance** *$ds = |d\mathbf{x}|$ and* **oriented volumes** *dV_k is defined by*

$$d\boldsymbol{x}^2 = \frac{1}{2}(\gamma_\mu\gamma_\nu + \gamma_\nu\gamma_\mu)dx^\mu dx^\nu = g_{\mu\nu}dx^\mu dx^\nu = \eta_{ab}\delta X^a \delta X^b, \tag{13}$$

$$dV_k = d\boldsymbol{x}_1 \wedge d\boldsymbol{x}_2 \wedge \cdots \wedge d\boldsymbol{x}_k = \gamma_{\mu\nu\cdots\omega} dx_1^\mu dx_2^\nu \cdots dx_k^\omega, \quad (1 \le k \le n), \tag{14}$$

in which $(\eta_{ab}) = diag\,(I_p, -I_q)$ is **Minkowski metric** *and $g_{\mu\nu}$ is* **Riemann metric**.

$$\gamma_{\mu\nu\cdots\omega} = \gamma_\mu \wedge \gamma_\nu \wedge \cdots \wedge \gamma_\omega \in \Lambda^k(T\mathbb{M}^{p,q})$$

is **Grassmann basis**. *The following Clifford-Grassmann number with basis*

$$c = c_0 I + c_\mu \gamma^\mu + c_{\mu\nu}\gamma^{\mu\nu} + \cdots + c_{12\cdots n}\gamma^{12\cdots n}, \quad (\forall c_k(x) \in \mathbb{R}) \tag{15}$$

defines real universal **Clifford algebra** *$C\ell_{p,q}$ on the manifold.*

The definitions and treatments in this chapter make the corresponding subtle and fallible concepts in differential geometry much simpler. For example, in spherical coordinate system of $\mathbb{R}^3$, we have element dx and the area element ds in sphere $dr = 0$ as

$$d\mathbf{x} = \sigma_1 dr + \sigma_2 r d\theta + \sigma_3 r \sin\theta d\varphi,$$

$$d\mathbf{s} = \sigma_2 r d\theta \wedge \sigma_3 r \sin\theta = i\sigma_1 r^2 \sin\theta d\theta d\varphi.$$

We have the total area of the sphere

$$\mathbf{A} = \oint d\mathbf{s} = i\sigma_1 r^2 \oint \sin\theta d\theta d\varphi = i\sigma_1 4\pi r^2.$$

The above definition involves a number of concepts, some more explanations are given in the following:

1. The geometrical meanings of elements $d\mathbf{x}, d\mathbf{y}, d\mathbf{x} \wedge d\mathbf{y}$ are shown in **Figure 2**. The relation between metric and vector basis is given by:

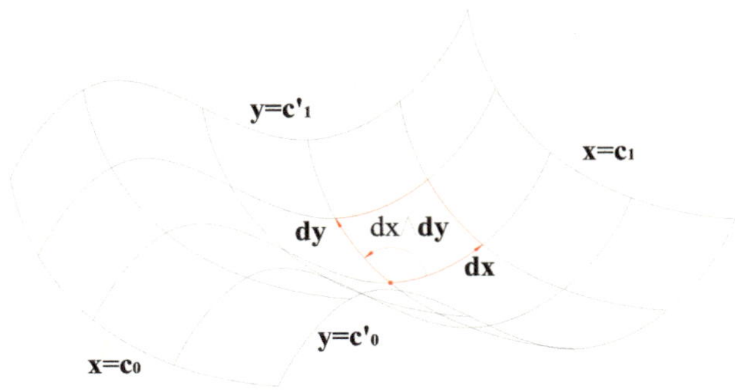

Figure 2.
Geometric meaning of vectors $\mathbf{dx}, \mathbf{dy}$ *and* $\mathbf{dx} \wedge \mathbf{dy}$.

$$g_{\mu\nu} = \frac{1}{2}(\gamma_\mu\gamma_\nu + \gamma_\nu\gamma_\mu) = \gamma_\mu \cdot \gamma_\nu, \tag{16}$$

$$\eta_{ab} = \frac{1}{2}(\gamma_a\gamma_b + \gamma_b\gamma_a) = \gamma_a \cdot \gamma_b, \tag{17}$$

which is the most important relation in Clifford algebra. Since Clifford algebra is isomorphic to some matrix algebra, by (17) γ_a is equivalent to some special matrices [20]. In practical calculation, we need not distinguish the vector basis from its representation matrix. The relation between the local **frame coefficient** $\left(f^\mu_a, f^a_\mu\right)$ and metric is given by:

$$\gamma^\mu = f^\mu_a\gamma^a, \quad \gamma_\mu = f^a_\mu\gamma_a, \quad f^a_\mu f^\mu_b = \delta^a_b, \quad f^a_\mu f^\nu_a = \delta^\nu_\mu.$$

$$f^\mu_a f^\nu_b\eta^{ab} = g^{\mu\nu}, \quad f^a_\mu f^b_\nu\eta_{ab} = g_{\mu\nu}.$$

2. Assume $\{\gamma_a | a = 1, 2\cdots n\}$ to be the basis of the space-time, then their exterior product is defined by [22]:

$$\gamma_{a_1} \wedge \gamma_{a_2} \cdots \wedge \gamma_{a_k} \equiv \frac{1}{k!}\sum_{\forall\sigma}\sigma^{b_1b_2\cdots b_k}_{a_1a_2\cdots a_k}\gamma_{b_1}\gamma_{b_2}\cdots\gamma_{b_k}, (1 \le k \le n).$$

In which $\sigma^{b_1b_2\cdots b_k}_{a_1a_2\cdots a_k}$ is permutation function, if $b_1b_2\cdots b_k$ is the even permutation of $a_1a_2\cdots a_k$, it equals 1. Otherwise, it equals -1. The above formula is a summation for all permutations, that is, it is antisymmetrization with respect to all indices. The geometric meaning of the exterior product is oriented volume of a higher dimensional parallel polyhedron. Exterior algebra is also called Grassmann algebra, which is associative.

3. By (12) and (13) we find that, using Clifford algebra to deal with the problems on a manifold or in the tangent space, the method is the same. Unless especially mentioned, we always use the Greek alphabet to stand for the index in curved space-time, and the Latin alphabet for the index in tangent space. We use Einstein summation convention.

4. In Eq. (15), each grade-k term is a tensor. For example, $c_0 I \in \Lambda^0$ is a scalar, $c_\mu\gamma^\mu \in \Lambda^1$ is a true vector, and $c_{\mu\nu}\gamma^{\mu\nu} \in \Lambda^2$ is an antisymmetric tensor of rank-2, which is also called a bivector, and so on. In practical calculation, coefficient

and basis should be written together, because they are one entity, such as (12) and (15). In this form, the variables become coordinate free. The coefficient is the value of tensor, which is just a number table, but the geometric meaning and transformation law of the tensor is carried by basis.

The real difficulty in learning modern mathematics is that in order to get a little result, we need a long list of subtle concepts. Mathematicians are used to defining concepts over concepts, but if the chain of concepts breaks down, the subsequent contents will not be understandable. Except for the professionals, the common readers impossibly have so much time to check and understand all concepts carefully. Fortunately, the Clifford algebra can avoid this problem, because Clifford algebra depends only on a few simple concepts, such as numbers, vectors, derivatives, and so on. The only somewhat new concept is the Clifford product of the vector bases, which is isomorphic to some special matrix algebra; and the rules of Clifford algebra are also standardized and suitable for brainless operations, which can be well mastered by high school students.

Definition 5 *For vector* $\boldsymbol{x} = \gamma_\mu x^\mu \in \Lambda^1$ *and multivector* $\boldsymbol{m} = \gamma_{\theta_1\theta_2\cdots\theta_k} m^{\theta_1\theta_2\cdots\theta_k} \in \Lambda^k$, *their inner product is defined as*

$$\boldsymbol{x} \odot \boldsymbol{m} = (\gamma_\mu \odot \gamma_{\theta_1\theta_2\cdots\theta_k}) x^\mu m^{\theta_1\theta_2\cdots\theta_k}, \qquad \boldsymbol{m} \odot \boldsymbol{x} = (\gamma_{\theta_1\theta_2\cdots\theta_k} \odot \gamma_\mu) x^\mu m^{\theta_1\theta_2\cdots\theta_k}, \quad (18)$$

in which

$$\gamma^\mu \odot \gamma^{\theta_1\theta_2\cdots\theta_k} \equiv g^{\mu\theta_1}\gamma^{\theta_2\cdots\theta_k} - g^{\mu\theta_2}\gamma^{\theta_1\theta_3\cdots\theta_k} + \cdots + (-1)^{k+1} g^{\mu\theta_k}\gamma^{\theta_1\cdots\theta_{k-1}}, \quad (19)$$

$$\gamma^{\theta_1\theta_2\cdots\theta_k} \odot \gamma^\mu \equiv (-1)^{k+1} g^{\mu\theta_1}\gamma^{\theta_2\cdots\theta_k} + (-1)^k g^{\mu\theta_2}\gamma^{\theta_1\theta_3\cdots\theta_k} + \cdots + g^{\mu\theta_k}\gamma^{\theta_1\cdots\theta_{k-1}}. \quad (20)$$

Theorem 2 *For basis of Clifford algebra, we have the following relations*

$$\gamma^\mu \gamma^{\theta_1\theta_2\cdots\theta_k} = \gamma^\mu \odot \gamma^{\theta_1\theta_2\cdots\theta_k} + \gamma^{\mu\theta_1\cdots\theta_k}, \quad (21)$$

$$\gamma^{\theta_1\theta_2\cdots\theta_k}\gamma^\mu = \gamma^{\theta_1\theta_2\cdots\theta_k} \odot \gamma^\mu + \gamma^{\theta_1\cdots\theta_k\mu}. \quad (22)$$

$$\gamma_{a_1a_2\cdots a_{n-1}} = \epsilon_{a_1a_2\cdots a_n}\gamma_{12\cdots n}\gamma^{a_n}, \quad (23)$$

$$\gamma_{a_1a_2\cdots a_{n-2}} = \frac{1}{2!}\epsilon_{a_1a_2\cdots a_n}\gamma_{12\cdots n}\gamma^{a_{n-1}a_n}, \quad (24)$$

$$\gamma_{a_1a_2\cdots a_{n-k}} = \frac{1}{k!}\epsilon_{a_1a_2\cdots a_n}\gamma_{12\cdots n}\gamma^{a_{n-k+1}\cdots a_n}. \quad (25)$$

Proof. Clearly $\gamma^\mu\gamma^{\theta_1\theta_2\cdots\theta_k} \in \Lambda^{k-1} \cup \Lambda^{k+1}$, so we have

$$\gamma^\mu\gamma^{\theta_1\theta_2\cdots\theta_k} = a_1 g^{\mu\theta_1}\gamma^{\theta_2\cdots\theta_k} + a_2 g^{\mu\theta_2}\gamma^{\theta_1\theta_3\cdots\theta_k} + \cdots + a_k g^{\mu\theta_k}\gamma^{\theta_1\cdots\theta_{k-1}} + A\gamma^{\mu\theta_1\cdots\theta_k}. \quad (26)$$

Permuting the indices θ_1 and θ_2, we find $a_2 = -a_1$. Let $\mu = \theta_1$, we get $a_1 = 1$. Check the monomial in exterior product, we get $A = 1$. Thus, we prove (21). In like manner, we prove (22). For orthonormal basis γ_a, by (22) we have:

$$\gamma_{a_1a_2\cdots a_{n-1}}\gamma_{a_n} = \epsilon_{a_1a_2\cdots a_n}\gamma_{12\cdots n}. \quad (27)$$

Again by $\gamma_{a_n}\gamma^{a_n} = 1$ (not summation), we prove (23). Other equations can be proved by antisymmetrization of indices. The proof is finished.

Likewise, we can define **multi-inner product** $\mathbf{A} \odot^k \mathbf{B}$ between multivectors as follows:

$$\gamma^{\mu\nu} \odot \gamma^{\alpha\beta} = g^{\mu\beta}\gamma^{\nu\alpha} - g^{\mu\alpha}\gamma^{\nu\beta} + g^{\nu\alpha}\gamma^{\mu\beta} - g^{\nu\beta}\gamma^{\mu\alpha}, \tag{28}$$

$$\gamma^{\mu\nu} \odot^2 \gamma^{\alpha\beta} = g^{\mu\beta}g^{\nu\alpha} - g^{\mu\alpha}g^{\nu\beta}, \quad \gamma^{\mu\nu} \odot^k \gamma^{\alpha\beta} = 0, (k > 2). \quad \cdots \tag{29}$$

We use $\mathbf{A} \odot^k \mathbf{B}$ rather $\mathbf{A}\cdot^k\mathbf{B}$, because the symbol "·" is too small to express exponential power. Then for the case $\gamma^{\mu_1\mu_2\cdots\mu_f}\gamma^{\theta_1\theta_2\cdots\theta_k}$, we have similar results. For example, we have

$$\gamma^{\mu\nu}\gamma^{\alpha\beta} = \gamma^{\mu\nu} \odot^2 \gamma^{\alpha\beta} + \gamma^{\mu\nu} \odot \gamma^{\alpha\beta} + \gamma^{\mu\nu\alpha\beta}. \tag{30}$$

In $C\ell_{1,3}$, denote the Pauli matrices by

$$\sigma^a \equiv \left\{ \begin{pmatrix} 1 & 0 \\ 0 & 1 \end{pmatrix}, \begin{pmatrix} 0 & 1 \\ 1 & 0 \end{pmatrix}, \begin{pmatrix} 0 & -i \\ i & 0 \end{pmatrix}, \begin{pmatrix} 1 & 0 \\ 0 & -1 \end{pmatrix} \right\}, \tag{31}$$

$$\sigma^0 = \tilde{\sigma}^0 = I, \qquad \tilde{\sigma}^k = -\sigma^k, \ (k = 1, 2, 3). \tag{32}$$

We use k, f, j standing for spatial indices. Define Dirac $\gamma-$ matrix by:

$$\gamma^a = \begin{pmatrix} 0 & \tilde{\sigma}^a \\ \sigma^a & 0 \end{pmatrix}, \qquad \gamma^5 = diag(I, -I). \tag{33}$$

γ^a forms the grade-1 basis of Clifford algebra $C\ell_{1,3}$. In equivalent sense, the representation (33) is unique. By γ-matrix (33), we have the complete bases of $C\ell_{1,3}$ as follows [21]:

$$I, \quad \gamma^a, \quad \gamma^{ab} = \frac{i}{2}\epsilon^{abcd}\gamma_{cd}\gamma^5, \quad \gamma^{abc} = i\epsilon^{abcd}\gamma_d\gamma^5, \quad \gamma^{0123} = -i\gamma^5. \tag{34}$$

Based on the above preliminaries, we can display some enlightening examples of application, which show how geometric algebra works efficiently. For a skew-symmetrical torsion $\mathcal{T}_{\mu\nu\omega} \equiv g_{\mu\beta}\mathcal{T}^{\beta}_{\ \nu\omega}$ in $\mathbb{M}^{1,3}$, by Clifford calculus, we have:

$$\mathcal{T} = \mathcal{T}_{\mu\nu\omega}\gamma^{\mu\nu\omega} = \mathcal{T}_{abc}\gamma^{abc} = \mathcal{T}_{abc}\epsilon^{abcd}\gamma_d(i\gamma^5) \equiv i\gamma_d\gamma^5\mathcal{T}^d = i\gamma_\alpha\gamma^5\mathcal{T}^\alpha, \tag{35}$$

and then

$$\mathcal{T}^\alpha = f^\alpha_d\mathcal{T}_{abc}\epsilon^{abcd} = \mathcal{T}_{\mu\nu\omega}f^\mu_a f^\nu_b f^\omega_c f^\alpha_d\epsilon^{abcd} = \frac{1}{\sqrt{g}}\epsilon^{\mu\nu\omega\alpha}\mathcal{T}_{\mu\nu\omega}, \tag{36}$$

where $g = |\det(g_{\mu\nu})|$. So we get:

$$\mathcal{T}_{\mu\nu\omega} = \sqrt{g}\epsilon_{\mu\nu\omega\alpha}\mathcal{T}^\alpha, \qquad \mathcal{T}_{\mu\nu\omega}\mathcal{T}^\omega = 0, \qquad \mathcal{T}^\alpha_{\mu\nu}\mathcal{T}^\nu = 0. \tag{37}$$

So, the skew-symmetrical torsion is equivalent to a pseudo vector in $\mathbb{M}^{1,3}$. This example shows the advantages to combine variable with basis together.

The following example discusses the absolute differential of tensors. The definition of vector, tensor, and spinor in differential geometry involving a number of refined concepts such as vector bundle and dual bundle, which are too complicated for readers in other specialty. Here, we inherit the traditional definitions based on the bases γ^a and γ^μ. In physics, basis of tensors is defined by direct products of grade-1 bases γ^μ. For metric, we have [23]:

$$\begin{aligned}\mathbf{g} &= g_{\mu\nu}\gamma^\mu \otimes \gamma^\nu = g^{\mu\nu}\gamma_\mu \otimes \gamma_\nu = \delta^\nu_\mu \gamma^\mu \otimes \gamma_\nu \\ &= \eta_{ab}\gamma^a \otimes \gamma^b = \eta^{ab}\gamma_a \otimes \gamma_b = \delta^b_a \gamma^a \otimes \gamma_b.\end{aligned} \tag{38}$$

For simplicity, we denote tensor basis by:

$$\otimes\gamma^{\mu_1\mu_2\cdots\mu_n} = \gamma^{\mu_1} \otimes \gamma^{\mu_2} \otimes \cdots \gamma^{\mu_n}, \quad \otimes\gamma_{\mu_1}{}^{\mu_2\mu_3\cdots\mu_n} = \gamma_{\mu_1} \otimes \gamma^{\mu_2} \otimes \cdots \gamma^{\mu_n}, \quad \cdots \tag{39}$$

In general, a tensor of rank n is given by:

$$\mathbf{T} = T_{\mu_1\mu_2\cdots\mu_n} \otimes \gamma^{\mu_1\mu_2\cdots\mu_n} = T^{\mu_1}{}_{\mu_2\cdots\mu_n} \otimes \gamma_{\mu_1}{}^{\mu_2\mu_3\cdots\mu_n} = \cdots \tag{40}$$

The geometrical information of the tensor such as transformation law and differential connection are all recorded by basis γ^μ, and all representations of rank (r, s) tensor denote the same one practical entity $\mathbf{T}(\mathbf{x})$. $T^{\nu\cdots}_{\mu\cdots}$ is just a quantity table similar to $c_{\mu\nu}$ in (15), but the physical and geometrical meanings of the tensor $\mathbf{T}$ are represented by basis γ^μ. Clifford algebra is a special kind of tensor with exterior product. Its algebraic calculus exactly reflects the intrinsic property of space-time and makes physical calculation simple and clear.

For the absolute differential of vector field $\mathbf{A} = \gamma_\mu A^\mu$, we have

$$\begin{aligned}dA &\equiv \lim_{\Delta x \to 0}[A(x + \Delta x) - A(x)] \\ &= (\partial_\alpha A^\mu \gamma_\mu + A^\mu d_\alpha \gamma_\mu) dx^\alpha = (\partial_\alpha A_\mu \gamma^\mu + A_\mu d_\alpha \gamma^\mu) dx^\alpha.\end{aligned} \tag{41}$$

We call $\eth_\alpha$ **connection operator** [23]. According to its geometrical meanings, connection operator should satisfy the following conditions:

1. It is a real linear transformation of basis γ^μ,

2. It satisfies metric consistent condition $d\mathbf{g} = 0$.

Thus, the differential connection can be generally expressed as:

$$\eth_\alpha \gamma^\mu = -\left(\Pi^\mu_{\alpha\beta} + T^{\;\mu}_{\alpha\beta}\right)\gamma^\beta, \qquad \Pi^\mu_{\alpha\beta} = \Pi^\mu_{\beta\alpha}, \qquad T^{\;\mu}_{\alpha\beta} = -T^{\;\mu}_{\beta\alpha}. \tag{42}$$

For metric $\mathbf{g} = g_{\mu\nu}\gamma^\mu \otimes \gamma^\nu$, by metric consistent condition we have:

$$\begin{aligned}0 &= d\mathbf{g} = d\left(g_{\mu\nu}\gamma^\mu \otimes \gamma^\nu\right) \\ &= \left[\left(\partial_\alpha g_{\mu\nu}\right)\gamma^\mu \otimes \gamma^\nu + g_{\mu\nu}(\eth_\alpha \gamma^\mu) \otimes \gamma^\nu + g_{\mu\nu}\gamma^\mu \otimes \eth_\alpha \gamma^\nu\right] dx^\alpha \\ &= \left[\left(\partial_\alpha g_{\mu\nu} - g_{\nu\beta}\Pi^\beta_{\alpha\mu} - g_{\mu\beta}\Pi^\beta_{\alpha\nu}\right)dx^\alpha - \left(g_{\nu\beta}T^{\;\beta}_{\alpha\mu} + g_{\mu\beta}T^{\;\beta}_{\alpha\nu}\right)dx^\alpha\right]\gamma^\mu \otimes \gamma^\nu.\end{aligned} \tag{43}$$

By (43), we have:

$$\left(\partial_\alpha g_{\mu\nu} - g_{\nu\beta}\Pi^\beta_{\alpha\mu} - g_{\mu\beta}\Pi^\beta_{\alpha\nu}\right)dx^\alpha - \left(g_{\nu\beta}T^{\;\beta}_{\alpha\mu} + g_{\mu\beta}T^{\;\beta}_{\alpha\nu}\right)dx^\alpha = 0. \tag{44}$$

Since $dx^\alpha \leftrightarrow \delta X^a$ is an arbitrary vector in tangent space, (44) is equivalent to:

$$\partial_\alpha g_{\mu\nu} - g_{\nu\beta}\Pi^\beta_{\alpha\mu} - g_{\mu\beta}\Pi^\beta_{\alpha\nu} = g_{\nu\beta}T^\beta_{\alpha\mu} + g_{\mu\beta}T^\beta_{\alpha\nu}. \tag{45}$$

(45) is a linear nonhomogeneous algebraic equation of $\left(\Pi^\mu_{\alpha\beta}, T^\mu_{\alpha\beta}\right)$.

Solving (45), we get the symmetrical particular solution "Christoffel symbols" as follows;

$$\Pi^\alpha_{\mu\nu} = \frac{1}{2}g^{\alpha\beta}\left(\partial_\mu g_{\beta\nu} + \partial_\nu g_{\mu\beta} - \partial_\beta g_{\mu\nu}\right) + \pi^\alpha_{\mu\nu} = \Gamma^\alpha_{\mu\nu} + \pi^\alpha_{\mu\nu}, \tag{46}$$

in which $\Gamma^\alpha_{\mu\nu}$ is called Levi-Civita connection determined by metric, $\pi^\alpha_{\mu\nu} = \pi^\alpha_{\nu\mu}$ is a symmetrical post-metric part of connection. In this chapter, the "**post-metric connection**" means the parts of connection cannot be determined by metric, i.e., the components $\pi^\alpha_{\mu\nu}$ and $T^\alpha_{\mu\nu}$ different from Levi-Civita connection $\Gamma^\alpha_{\mu\nu}$. Denote

$$\mathcal{T}_{\mu|\nu\alpha} = g_{\mu\beta}T^\beta_{\nu\alpha}, \quad \pi_{\mu|\nu\alpha} = g_{\mu\beta}\pi^\beta_{\nu\alpha}, \quad K_{\mu\nu\alpha} = \pi_{\mu|\nu\alpha} + \mathcal{T}_{\mu|\nu\alpha}, \tag{47}$$

where $K_{\mu\nu\alpha}$ is called contortion with total n^3 components [24]. Substituting (46) and (47) into metric compatible condition (45), we get $\frac{1}{2}(n+1)n^2$ constraints for $K_{\mu\nu\alpha}$,

$$K_{\mu\nu\alpha} + K_{\nu\mu\alpha} = 0 = \left(\pi_{\mu|\nu\alpha} + \pi_{\nu|\mu\alpha}\right) + \left(\mathcal{T}_{\mu|\nu\alpha} + \mathcal{T}_{\nu|\mu\alpha}\right). \tag{48}$$

By (48), $K_{\mu\nu\alpha}$ has only $\frac{1}{2}(n-1)n^2$ independent components. Noticing torsion $\mathcal{T}_{\mu|\nu\alpha}$ has just $\frac{1}{2}(n-1)n^2$ independent components, so $K_{\mu\nu\alpha}$ or $\pi_{\mu|\nu\alpha}$ can be represented by $\mathcal{T}_{\mu|\nu\alpha}$.

Theorem 3 *For post-metric connections we have the following relations*

$$\pi_{\mu|\nu\alpha} = \mathcal{T}_{\nu|\alpha\mu} + \mathcal{T}_{\alpha|\nu\mu}, \tag{49}$$

$$K_{\mu\nu\alpha} = \mathcal{T}_{\nu|\alpha\mu} + \mathcal{T}_{\alpha|\nu\mu} + \mathcal{T}_{\mu|\nu\alpha}, \tag{50}$$

$$\mathcal{T}_{\mu|\nu\alpha} = \frac{1}{3}\left(\pi_{\alpha|\mu\nu} - \pi_{\nu|\mu\alpha}\right) + \tilde{\mathcal{T}}_{\mu\nu\alpha}, \tag{51}$$

and consistent condition

$$\pi_{\mu|\nu\alpha} + \pi_{\alpha|\mu\nu} + \pi_{\nu|\alpha\mu} = 0. \tag{52}$$

$\tilde{\mathcal{T}} = \tilde{\mathcal{T}}_{\mu\nu\omega}\gamma^{\mu\nu\omega} \in \Lambda^3$ *is an arbitrary skew-symmetrical tensor*.

Proof If we represent $\pi_{\mu|\nu\alpha}$ by $\mathcal{T}_{\mu|\nu\alpha}$, by (48) and symmetry we have solution as (49). By (49), we get consistent condition (52). By (49) and (47), we get (50).

If we represent $\mathcal{T}_{\mu|\nu\alpha}$ by $\pi_{\mu|\nu\alpha}$, we generally have linear relation

$$\mathcal{T}_{\mu|\nu\alpha} = k\left(\pi_{\nu|\mu\alpha} - \pi_{\alpha|\mu\nu}\right) + \tilde{\mathcal{T}}_{\mu\nu\alpha}, \tag{53}$$

in which k is a constant to be determined, $\tilde{\mathcal{T}}_{\mu\nu\alpha}$ is particular solution as $\pi_{\mu|\nu\alpha} \equiv 0$. $\tilde{\mathcal{T}}_{\mu\nu\alpha}$ satisfies

$$\tilde{\mathcal{T}}_{\mu\nu\alpha} = \tilde{\mathcal{T}}_{\alpha\mu\nu} = \tilde{\mathcal{T}}_{\nu\alpha\mu} = -\tilde{\mathcal{T}}_{\mu\alpha\nu} = -\tilde{\mathcal{T}}_{\nu\mu\alpha} = -\tilde{\mathcal{T}}_{\alpha\nu\mu}. \tag{54}$$

So this part of torsion is a skew-symmetrical tensor $\tilde{\mathcal{T}} = \tilde{\mathcal{T}}_{\mu\nu\omega}\gamma^{\mu\nu\omega} \in \Lambda^3$, which has $C^3_n = \frac{1}{6}(n-2)(n-1)n$ independent components. Substituting (53) into (48), we get

$$(k-1)\left(\pi_{\mu|\nu\alpha}+\pi_{\nu|\mu\alpha}\right)=2k\pi_{\alpha|\mu\nu}. \tag{55}$$

Calculating the summation of (55) for circulation of $\{\mu,\nu,\alpha\}$, we also get consistent condition (52). Substituting (52) into (55) we get $k=\frac{1}{3}$. Again by (53), we get solution (51). It is easy to check, (49) and (51) are the inverse representation under condition (52). The proof is finished.

Substituting (42) into

$$0=d\mathbf{g}=\delta_\nu^\mu\left[\left(\partial_\alpha\gamma_\mu\right)\otimes\gamma^\nu+\gamma_\mu\otimes\partial_\alpha\gamma^\nu\right]dx^\alpha, \tag{56}$$

we get

$$\partial_\alpha\gamma_\mu=\left(\Gamma_{\alpha\mu}^\nu+\pi_{\alpha\mu}^\nu+\mathcal{T}_{\alpha\mu}^\nu\right)\gamma_\nu. \tag{57}$$

To understand the meaning of $\pi_{\mu\nu}^\alpha$ and $\mathcal{T}_{\mu\nu}^\alpha$, we examine the influence on geodesic.

$$\begin{aligned}\frac{d\mathbf{v}}{ds}\equiv\frac{dv^\alpha}{ds}\gamma_\alpha+v^\alpha\partial_\mu\gamma_\alpha v^\mu&=\left(\frac{dv^\alpha}{ds}+\left(\Gamma_{\mu\nu}^\alpha+\pi_{\mu\nu}^\alpha+\mathcal{T}_{\mu\nu}^\alpha\right)v^\mu v^\nu\right)\gamma_\alpha,\\&=\left(\frac{d}{ds}v^\alpha+\Gamma_{\mu\nu}^\alpha v^\mu v^\nu\right)\gamma_\alpha+\pi_{\mu\nu}^\alpha v^\mu v^\nu\gamma_\alpha.\end{aligned} \tag{58}$$

The term $\mathcal{T}_{\mu\nu}^\alpha v^\mu v^\nu=0$ due to $\mathcal{T}_{\mu\nu}^\alpha=-\mathcal{T}_{\nu\mu}^\alpha$. So the symmetrical part $\pi_{\mu\nu}^\alpha$ influences the geodesic, but the antisymmetrical part $\mathcal{T}_{\mu\nu}^\alpha$ only influences spin of a particle. This means $\pi_{\mu\nu}^\alpha\neq 0$ violates Einstein's equivalent principle. In what follows, we take $\pi_{\mu\nu}^\alpha=0$.

By (42) and (57), we get:

Theorem 4 *In the case* $\pi_{\mu\nu}^\alpha\equiv 0$, *the absolute differential of vector* $\boldsymbol{A}$ *is given by*

$$d\boldsymbol{A}=\nabla_\alpha A^\mu\gamma_\mu dx^\alpha=\nabla_\alpha A_\mu\gamma^\mu dx^\alpha, \tag{59}$$

in which ∇_α *denotes the absolute derivatives of vector defined as follows:*

$$\nabla_\alpha A^\mu=A^\mu_{;\alpha}+\mathcal{T}^\mu_{\alpha\beta}A^\beta,\quad A^\mu_{;\alpha}=\partial_\alpha A^\mu+\Gamma^\mu_{\alpha\nu}A^\nu, \tag{60}$$

$$\nabla_\alpha A_\mu=A_{\mu;\alpha}-\mathcal{T}^\beta_{\alpha\mu}A_\beta,\quad A_{\mu;\alpha}=\partial_\alpha A_\mu-\Gamma^\nu_{\alpha\mu}A_\nu, \tag{61}$$

where $A^\mu_{;\alpha}$ and $A_{\mu;\alpha}$ *are usual covariant derivatives of vector without torsion. Torsion* $\mathcal{T}_{\mu\nu\omega}\in\Lambda^3$ *is an antisymmetrical tensor of* C_n^3 *independent components.*

Similarly, we can calculate the absolute differential for any tensor. The example also shows the advantages to combine variable with basis.

Now we take spinor connection as example to show the power of Clifford algebra. For Dirac equation in curved space-time without torsion, we have [23, 25, 26]:

$$\gamma^\mu i(\partial_\mu+\Gamma_\mu)\phi=m\phi,\qquad\Gamma_\mu=\frac{1}{4}\gamma_\nu\left(\partial_\mu\gamma^\nu+\Gamma^\nu_{\mu\alpha}\gamma^\alpha\right). \tag{62}$$

Γ_μ is called spinor connection. Representing $\gamma^\mu\Gamma_\mu\in\Lambda^1\cup\Lambda^3$ in the form of (15), we get:

$$\alpha^\mu\hat{p}_\mu\phi-s_\mu\Omega^\mu\phi=m\gamma^0\phi, \tag{63}$$

where α^μ is current operator, $\hat{p}_\mu$ is momentum operator, and s_μ spin operator. They are defined respectively by:

$$\alpha^\mu = \operatorname{diag}(\sigma^\mu, \tilde{\sigma}^\mu), \quad \hat{p}_\mu = i(\partial_\mu + \Upsilon_\mu) - eA_\mu, \quad s^\mu = \frac{1}{2}\operatorname{diag}(\sigma^\mu, -\tilde{\sigma}^\mu), \tag{64}$$

where $\sigma^\mu = f^\mu_a \sigma^a$ and $\tilde{\sigma}^\mu = f^\mu_a \tilde{\sigma}^a$ are the Pauli matrices in curved space-time. $\Upsilon_\mu \in \Lambda^1$ is called Keller connection, and $\Omega_\mu \in \Lambda^3$ is called Gu-Nester potential, which is a pseudo vector [23, 26, 27]. They are calculated by:

$$\Upsilon_\mu = \frac{1}{2} f^\nu_a \left(\partial_\mu f^a_\nu - \partial_\nu f^a_\mu\right), \quad \Omega^\alpha = \frac{1}{2}\epsilon^{abcd} f^\alpha_d f^\mu_a f^\nu_b \partial_\mu f^e_\nu \eta_{ce} = \frac{1}{4}\epsilon^{dabc} f^\alpha_d f^\beta_a S^{\mu\nu}_{bc} \partial_\beta g_{\mu\nu}, \tag{65}$$

where $S^{\mu\nu}_{ab} \equiv f^{\{\mu}_a f^{\nu\}}_b \operatorname{sign}(a-b)$ for LU decomposition of metric. In the Hamiltonian of a spinor, we get a spin-gravity coupling potential $s_\mu \Omega^\mu$. If the metric of the space-time can be orthogonalized, we have $\Omega_\mu \equiv 0$.

If the gravitational field is generated by a rotating ball, the corresponding metric, like the Kerr one, cannot be diagonalized. In this case, the spin-gravity coupling term has nonzero coupling effect. In asymptotically flat space-time, we have the line element in quasi-spherical coordinate system [28]:

$$dx = \gamma_0 \sqrt{U}(dt + Wd\varphi) + \sqrt{V}(\gamma_1 dr + \gamma_2 r d\theta) + \gamma_3 \sqrt{U^{-1}} r \sin\theta d\varphi, \tag{66}$$

$$dx^2 = U(dt + Wd\varphi)^2 - V\left(dr^2 + r^2 d\theta^2\right) - U^{-1} r^2 \sin^2\theta d\varphi^2, \tag{67}$$

in which (U, V, W) is just functions of (r, θ). As $r \to \infty$ we have:

$$U \to 1 - \frac{2m}{r}, \quad W \to \frac{4L}{r}\sin^2\theta, \quad V \to 1 + \frac{2m}{r}, \tag{68}$$

where (m, L) are mass and angular momentum of the star, respectively. For common stars and planets, we always have $r \gg m \gg L$. For example, we have $m \doteq 3$ km for the sun. The nonzero tetrad coefficients of metric (66) are given by:

$$\begin{cases} f^0_t = \sqrt{U}, f^1_r = \sqrt{V}, f^2_\theta = r\sqrt{V}, f^3_\varphi = \dfrac{r\sin\theta}{\sqrt{U}}, f^0_\varphi = \sqrt{U}W, \\ f^t_0 = \dfrac{1}{\sqrt{U}}, f^r_1 = \dfrac{1}{\sqrt{V}}, f^\theta_2 = \dfrac{1}{r\sqrt{V}}, f^\varphi_3 = \dfrac{\sqrt{U}}{r\sin\theta}, f^t_3 = \dfrac{-\sqrt{U}W}{r\sin\theta}. \end{cases} \tag{69}$$

Substituting it into (65) we get

$$\begin{aligned} \Omega^\alpha &= f^t_0 f^r_1 f^\theta_2 f^\varphi_3 \left(0, \partial_\theta g_{t\varphi}, -\partial_r g_{t\varphi}, 0\right) \\ &= \left(Vr^2 \sin\theta\right)^{-1} (0, \partial_\theta(UW), -\partial_r(UW), 0) \\ &\to \frac{4L}{r^4}(0, 2r\cos\theta, \sin\theta, 0). \end{aligned} \tag{70}$$

By (70), we find that the intensity of Ω^α is proportional to the angular momentum of the star, and its force line is given by:

$$\frac{dx^\mu}{ds} = \Omega^\mu \Rightarrow \frac{dr}{d\theta} = \frac{2r\cos\theta}{\sin\theta} \Leftrightarrow r = R\sin^2\theta. \tag{71}$$

(71) shows that the force lines of Ω^α is just the magnetic lines of a magnetic dipole. According to the above results, we know that the spin-gravity coupling potential of charged particles will certainly induce a macroscopic dipolar magnetic field for a star, and it should be approximately in accordance with the Schuster-Wilson-Blackett relation [29–31].

3. Representation of Clifford algebra

The matrix representation of Clifford algebra is an old problem with a long history. As early as in 1908, Cartan got the following periodicity of 8 [18, 19].

Theorem 5 *For real universal Clifford algebra $C\ell_{p,q}$, we have the following isomorphism*

$$C\ell_{p,q} \cong \begin{cases} \mathrm{Mat}(2^{\frac{n}{2}}, \mathbb{R}), & \text{if } \mathrm{mod}(p-q,8) = 0,2 \\ \mathrm{Mat}\left(2^{\frac{n-1}{2}}, \mathbb{R}\right) \oplus \mathrm{Mat}\left(2^{\frac{n-1}{2}}, \mathbb{R}\right), & \text{if } \mathrm{mod}(p-q,8) = 1 \\ \mathrm{Mat}\left(2^{\frac{n-1}{2}}, \mathbb{C}\right), & \text{if } \mathrm{mod}(p-q,8) = 3,7 \\ \mathrm{Mat}\left(2^{\frac{n-2}{2}}, \mathbb{H}\right), & \text{if } \mathrm{mod}(p-q,8) = 4,6 \\ \mathrm{Mat}\left(2^{\frac{n-3}{2}}, \mathbb{H}\right) \oplus \mathrm{Mat}\left(2^{\frac{n-3}{2}}, \mathbb{H}\right), & \text{if } \mathrm{mod}(p-q,8) = 5. \end{cases} \tag{72}$$

For $C\ell_{0,2}$, we have $C = tI + x\gamma_1 + y\gamma_2 + z\gamma_{12}$ with

$$\gamma_1^2 = \gamma_2^2 = \gamma_{12}^2 = -1, \gamma_1\gamma_2 = -\gamma_2\gamma_1 = \gamma_{12}, \gamma_2\gamma_{12} = -\gamma_{12}\gamma_2 = \gamma_1, \gamma_{12}\gamma_1 = -\gamma_1\gamma_{12} = \gamma_2. \tag{73}$$

By (73), we find C is equivalent to a quaternion, that is, we have isomorphic relation $C\ell_{0,2} \cong \mathbb{H}$.

Similarly, for $C\ell_{2,0}$, we have $C = tI + x\gamma_1 + y\gamma_2 + z\gamma_{12}$ with

$$\gamma_1^2 = \gamma_2^2 = \gamma_{12}^2 = 1, \gamma_1\gamma_2 = -\gamma_2\gamma_1 = \gamma_{12}, \gamma_2\gamma_{12} = -\gamma_{12}\gamma_2 = -\gamma_1, \gamma_{12}\gamma_1 = -\gamma_1\gamma_{12} = -\gamma_2. \tag{74}$$

By (74), the basis is equivalent to

$$\gamma_1 = \begin{pmatrix} 0 & 1 \\ 1 & 0 \end{pmatrix}, \quad \gamma_2 = \begin{pmatrix} 1 & 0 \\ 0 & -1 \end{pmatrix}, \quad \gamma_{12} = \begin{pmatrix} 0 & -1 \\ 1 & 0 \end{pmatrix}. \tag{75}$$

Thus, (75) means $C\ell_{2,0} \cong \mathrm{Mat}(2, \mathbb{R})$.

In geometry and physics, the matrix representation of generators of Clifford algebra is more important and fundamental than the representation of whole algebra. Define γ^μ by

$$\gamma^\mu = \begin{pmatrix} 0 & \tilde{\vartheta}^\mu \\ \vartheta^\mu & 0 \end{pmatrix} \equiv \Gamma^\mu(m), \quad \vartheta_\mu = \mathrm{diag}\left(\overbrace{\sigma_\mu, \sigma_\mu, \cdots, \sigma_\mu}^{m}\right), \quad \tilde{\vartheta}_\mu = \mathrm{diag}\left(\overbrace{\tilde{\sigma}_\mu, \tilde{\sigma}_\mu, \cdots, \tilde{\sigma}_\mu}^{m}\right). \tag{76}$$

which forms the generator or grade-1 basis of Clifford algebra $C\ell_{1,3}$. To denote γ^μ by $\Gamma^\mu(m)$ is for the convenience of representation of high dimensional Clifford algebra. For any matrices C^μ satisfying $C\ell_{1,3}$ Clifford algebra, we have [20, 32]:

Theorem 6 *Assuming the matrices C^μ satisfy anti-commutative relation of $C\ell_{1,3}$*

$$C^\mu C^\nu + C^\nu C^\mu = 2\eta^{\mu\nu}, \tag{77}$$

then there is a natural number m and an invertible matrix K, such that $K^{-1}C^\mu K = \Gamma^\mu(m)$.

This means in equivalent sense, we have unique representation (76) for generator of $C\ell_{1,3}$. In [20], we derived complex representation of generators of $C\ell_{p,q}$ based on Theorem 6 and real representations according to the complex representations as follows.

Theorem 7 *Let*

$$\gamma^5 = i\,diag\,(E, -E), \quad E \equiv diag\,(I_{2k}, -I_{2l}), \quad k + l = n. \tag{78}$$

Other γ^μ, $(\mu \le 3)$ are given by (76)*. Then the generators of Clifford algebra $C\ell_{1,4}$ are equivalent to $\forall\gamma^\mu$, $(\mu = 0, 1, 2, 3, 5)$.*

In order to express the general representation of generators, we introduce some simple notations. I_m stands for $m \times m$ unit matrix. For any matrix $A = (A_{ab})$, denote block matrix

$$A \otimes I_m = (A_{ab} I_m), \qquad [A, B, C, \cdots] = \mathrm{diag}\,(A, B, C, \cdots). \tag{79}$$

in which the direct product of matrix is Kronecker product. Obviously, we have $I_2 \otimes I_2 = I_4$, $I_2 \otimes I_2 \otimes I_2 = I_8$, and so on. In what follows, we use $\Gamma^\mu(m)$ defined in (76). For $\mu \in \{0, 1, 2, 3\}$, $\Gamma^\mu(m)$ is $4m \times 4m$ matrix, which constitute the generator of $C\ell_{1,3}$. Similar to the above proofs, we can check the following theorem by method of induction.

Theorem 8

1. *In equivalent sense, for $C\ell_{4m}$, the matrix representation of generators is uniquely given by*

$$\Big\{\Gamma^\mu(n),\ \Big[\Gamma^\mu\Big(\frac{n}{2^2}\Big), -\Gamma^\mu\Big(\frac{n}{2^2}\Big)\Big] \otimes I_2,$$
$$\Big[\Big[\Gamma^\mu\Big(\frac{n}{2^4}\Big), -\Gamma^\mu\Big(\frac{n}{2^4}\Big)\Big], -\Big[\Gamma^\mu\Big(\frac{n}{2^4}\Big), -\Gamma^\mu\Big(\frac{n}{2^4}\Big)\Big]\Big] \otimes I_{2^2},$$
$$\Big[\Gamma^\mu\Big(\frac{n}{2^6}\Big), -\Gamma^\mu\Big(\frac{n}{2^6}\Big), -\Gamma^\mu\Big(\frac{n}{2^6}\Big), \Gamma^\mu\Big(\frac{n}{2^6}\Big), -\Gamma^\mu\Big(\frac{n}{2^6}\Big), \Gamma^\mu\Big(\frac{n}{2^6}\Big), \Gamma^\mu\Big(\frac{n}{2^6}\Big), -\Gamma^\mu\Big(\frac{n}{2^6}\Big)\Big] \otimes I_{2^3}, \cdots\Big\}. \tag{80}$$

in which $n = 2^{m-1}N$, where N is any given positive integer. All matrices are $2^{m+1}N \times 2^{m+1}N$ type.

2. *For $C\ell_{4m+1}$, besides* (80) *we have another real generator*

$$\gamma^{4m+1} = [E, -E, -E, E, -E, E, E, -E\cdots], \qquad E = [I_{2k}, -I_{2l}]. \tag{81}$$

If and only if $k = l$, this representation can be uniquely expanded as generators of $C\ell_{4m+4}$.

3. *For any* $C\ell_{p,q}$, $\{p,q|p+q\leq 4m, \mathrm{mod}(p+q,4)\neq 1\}$, *the combination of* $p+q$ *linear independent generators* $\{\gamma^\mu, i\gamma^\nu\}$ *taking from* (80) *constitutes the complete set of generators. In the case* $\{p,q|p+q\leq 4m, \mathrm{mod}(p+q,4)=1\}$, *besides the combination of* $\{\gamma^\mu, i\gamma^\nu\}$, *we have another normal representation of generator taking the form* (81) *with* $k\neq l$.

4. *For* $C\ell_m$, $(m<4)$, *we have another* 2×2 *Pauli matrix representation for its generators* $\{\sigma^1,\sigma^2,\sigma^3\}$.

Then, we get all complex matrix representations for generators of real $C\ell_{p,q}$ explicitly.

The real representation of $C\ell_{p,q}$ can be easily constructed from the above complex representation. In order to get the real representation, we should classify the generators derived above. Let $\mathbf{G}_c(n)$ stand for any one set of all complex generators of $C\ell_n$ given in Theorem 8, and set the coefficients before all σ^μ and $\tilde{\sigma}^\mu$ as 1 or i. Denote $\mathbf{G}_{c+}$ stands for the set of complex generators of $C\ell_{n,0}$ and $\mathbf{G}_{c-}$ for the set of complex generators of $C\ell_{0,n}$. Then, we have:

$$\mathbf{G}_c = \mathbf{G}_{c+}\cup \mathbf{G}_{c-}, \qquad \mathbf{G}_{c-}\cong i\mathbf{G}_{c+}. \tag{82}$$

By the construction of generators, we have only two kinds of γ^μ matrices. One is the matrix with real nonzero elements and the other is that with imaginary nonzero elements. This is because all nonzero elements of σ^2 are imaginary but all other $\sigma^\mu(\forall\mu\neq 2)$ are real. Again assume

$$\mathbf{G}_{c+} = \mathbf{G}_r\cup\mathbf{G}_i, \qquad \mathbf{G}_r = \{\gamma_r^\mu|\gamma_r^\mu \text{ is real}\}, \qquad \mathbf{G}_i = \{\gamma_i^\mu|\gamma_i^\mu \text{ is imaginary}\}. \tag{83}$$

Denote $J_2 = i\sigma^2$, we have $J_2^2 = -I_2$. J_2 becomes the real matrix representation for imaginary unit i. Using the direct products of complex generators with (I_2, J_2), we can easily construct the real representation of all generators for $C\ell_{p,q}$ from $\mathbf{G}_{c+}$ as follows.

Theorem 9

1. *For* $C\ell_{n,0}$, *we have real matrix representation of generators as*

$$\mathbf{G}_{r+} = \{\gamma^\mu\otimes I_2\ (\text{if } \gamma^\mu\in\mathbf{G}_r);\ \ i\gamma^\nu\otimes J_2\ (\text{if } \gamma^\nu\in\mathbf{G}_i)\}. \tag{84}$$

2. *For* $C\ell_{0,n}$, *we have real matrix representation of generators as*

$$\mathbf{G}_{r-} = \{\gamma^\mu\otimes J_2|\gamma^\mu\in\mathbf{G}_{r+}\}. \tag{85}$$

3. *For* $C\ell_{p,q}$, *we have real matrix representation of generators as*

$$\mathbf{G}_r = \left\{\Gamma_+^{\mu_a}, \Gamma_-^{\nu_b} \left| \begin{matrix} \Gamma_+^{\mu_a} = \gamma^{\mu_a}\in\mathbf{G}_{r+}, (a=1,2,\cdots,p) \\ \Gamma_-^{\nu_b} = \gamma^{\nu_b}\in\mathbf{G}_{r-}, (b=1,2,\cdots,q) \end{matrix}\right.\right\}. \tag{86}$$

Obviously we have $C_n^pC_n^q = (C_n^p)^2$ *choices for the real generators of* $C\ell_{p,q}$ *from each complex representation*.

Proof. By calculating rules of block matrix, it is easy to check the following relations:

$$(\gamma^\mu\otimes I_2)(\gamma^\nu\otimes J_2) + (\gamma^\nu\otimes J_2)(\gamma^\mu\otimes I_2) = (\gamma^\mu\gamma^\nu + \gamma^\nu\gamma^\mu)\otimes J_2, \tag{87}$$

$$(\gamma^\mu \otimes J_2)(\gamma^\nu \otimes J_2) + (\gamma^\nu \otimes J_2)(\gamma^\mu \otimes J_2) = -(\gamma^\mu\gamma^\nu + \gamma^\nu\gamma^\mu) \otimes I_2. \tag{88}$$

By these relations, Theorem 9 becomes a direct result of Theorem 8.

For example, we have 4×4 real matrix representation for generators of $C\ell_{0,3}$ as follows:

$$i\{\sigma^1, \sigma^2, \sigma^3\} \cong \{\sigma^1 \otimes J_2, i\sigma^2 \otimes I_2, \sigma^3 \otimes J_2\} \equiv \{\Sigma^1, \Sigma^2, \Sigma^3\} =$$
$$\left\{ \begin{pmatrix} 0 & 0 & 0 & 1 \\ 0 & 0 & -1 & 0 \\ 0 & 1 & 0 & 0 \\ -1 & 0 & 0 & 0 \end{pmatrix}, \begin{pmatrix} 0 & 0 & 1 & 0 \\ 0 & 0 & 0 & 1 \\ -1 & 0 & 0 & 0 \\ 0 & -1 & 0 & 0 \end{pmatrix}, \begin{pmatrix} 0 & 1 & 0 & 0 \\ -1 & 0 & 0 & 0 \\ 0 & 0 & 0 & -1 \\ 0 & 0 & 1 & 0 \end{pmatrix} \right\}. \tag{89}$$

It is easy to check

$$\Sigma^k\Sigma^l + \Sigma^l\Sigma^k = -2\delta^{kl}, \qquad \Sigma^k\Sigma^l - \Sigma^l\Sigma^k = 2\epsilon^{klm}\Sigma_m. \tag{90}$$

4. Transformation of Clifford algebra

Assume V is the base vector space of $C\ell_{p,q}$, then Clifford algebra has the following global properties [22, 33, 34]:

$$C\ell_{p,q} = \bigoplus_{k=0}^{n} \Lambda^k V = C\ell^+_{p,q} \oplus C\ell^-_{p,q}, \tag{91}$$

$$C\ell^+_{p,q} \equiv \bigoplus_{k=\text{even}} \Lambda^k V, \quad C\ell^-_{p,q} \equiv \bigoplus_{k=\text{odd}} \Lambda^k V, \tag{92}$$

$$C\ell_{p,q} \cong C\ell^+_{p,q+1}. \tag{93}$$

$C\ell_{p,q}$ is a $\mathbb{Z}_2$-graded superalgebra, and $C\ell^+_{p,q}$ is a subalgebra of $C\ell_{p,q}$. We have:

$$C\ell^+C\ell^+ = C\ell^-C\ell^- = C\ell^+, \quad C\ell^+C\ell^- = C\ell^-C\ell^+ = C\ell^-. \tag{94}$$

Definition 6 *The* **conjugation** *of element in* $C\ell_{p,q}$ *is defined by*

$$\left(\gamma_{k_1k_2\cdots k_m}\right)^* = (-1)^m \gamma_{k_m\cdots k_2k_1} = (-1)^{\frac{1}{2}m(m+1)}\gamma_{k_1k_2\cdots k_m}, (0 \le m \le n). \tag{95}$$

The **main involution** *of element is defined by*

$$\alpha\left(\gamma_{k_1k_2\cdots k_m}\right) = (-1)^m \gamma_{k_1k_2\cdots k_m}, (0 \le m \le n). \tag{96}$$

The **norm** *and* **inverse** *of element are defined by*

$$\mathcal{N}(\boldsymbol{X}) \equiv \boldsymbol{X}\boldsymbol{X}^*, \quad \boldsymbol{X}^{-1} = \boldsymbol{X}^*/N(\boldsymbol{X}) \text{ if } \mathcal{N}(\boldsymbol{X}) \ne 0. \tag{97}$$

By the definition, it is easy to check

$$\gamma_k^* = -\gamma_k, \quad \gamma_{ab}^* = -\gamma_{ab}, \quad \gamma_{abc}^* = \gamma_{abc}, \quad \cdots \tag{98}$$

$$\alpha(\mathbf{x}^*) = \alpha(\mathbf{x})^*, \quad \alpha(\gamma_k) = -\gamma_k, \quad \alpha(\gamma_{ab}) = \gamma_{ab}, \quad \cdots \tag{99}$$

$$\mathbf{g}^{-1} = \mathbf{g}^*, \quad \{\mathbf{g} = \mathbf{g}_1\mathbf{g}_2\cdots\mathbf{g}_m | \forall \mathbf{g}_k \in \Lambda^1, N(\mathbf{g}_k) = 1\}. \tag{100}$$

Definition 7 *The* **Pin group** *and* **Spin group** *of* $C\ell_{p,q}$ *are defined by*

$$Pin_{p,q} = \{ \mathbf{g} \in C\ell_{p,q} | \mathcal{N}(\mathbf{g}) = \pm 1, \alpha(\mathbf{g})\mathbf{x}\mathbf{g}^* \in V \forall \mathbf{x} \in V \}, \tag{101}$$

$$Spin_{p,q} = \left\{ \mathbf{g} \in C\ell^+_{p,q} | \mathcal{N}(\mathbf{g}) = \pm 1, \mathbf{g}\mathbf{x}\mathbf{g}^* \in V \forall \mathbf{x} \in V \right\} = Pin \cap C\ell^+. \tag{102}$$

The transformation $\mathbf{x} \mapsto \alpha(\mathbf{g})\mathbf{x}\mathbf{g}^*$ is called sandwich operator. Pin or Spin group consists of two connected components with $\mathcal{N}(\mathbf{g}) = 1$ or $\mathcal{N}(\mathbf{g}) = -1$,

$$\text{Spin}^+_{p,q} = \left\{ \mathbf{g} \in C\ell^+_{p,q} | \mathcal{N}(\mathbf{g}) = +1, \mathbf{g}\mathbf{x}\mathbf{g}^* \in V \forall \mathbf{x} \in V \right\}, \tag{103}$$

$$\text{Spin}^-_{p,q} = \left\{ \mathbf{g} \in C\ell^+_{p,q} | \mathcal{N}(\mathbf{g}) = -1, \mathbf{g}\mathbf{x}\mathbf{g}^* \in V \forall \mathbf{x} \in V \right\}. \tag{104}$$

For $\forall \mathbf{g} \in \text{Pin}_{p,q}, \mathbf{x} \in V$, the sandwich operator is a linear transformation for vector in V,

$$\mathbf{x}' = \alpha(\mathbf{g})\mathbf{x}\mathbf{g}^* \Rightarrow X' = KX, X = \left(x^1, x^2 \cdots x^n\right)^T. \tag{105}$$

In all transformations of vector, the reflection and rotation transformations are important in geometry. Here, we discuss the transformation in detail. Let $\mathbf{m} \in \Lambda^1$ be a unit vector in V, then the **reflection transformation** of vector $\mathbf{X} \in \Lambda^1$ with respect to $n - 1$ dimensional mirror perpendicular to $\mathbf{m}$ is defined by [35]:

$$\mathbf{X}' = \mathbf{m}\mathbf{X}\mathbf{m}^* = -\mathbf{m}\mathbf{X}\mathbf{m}. \tag{106}$$

Let $\mathbf{m} = \gamma_a m^a, \mathbf{X} = \gamma_a X^a$, substituting it into (106) and using (21), we have:

$$\begin{aligned} \mathbf{X}' &= -(\mathbf{m} \odot \mathbf{X} + m^a X^b \gamma_{ab})\mathbf{m} = -(\mathbf{m} \odot \mathbf{X})\mathbf{m} - m^a X^b m^c (\gamma_{ab}\gamma_c) \\ &= -(\mathbf{m} \odot \mathbf{X})\mathbf{m} - m^a X^b m^c (g_{bc}\gamma_a - g_{ac}\gamma_b + \gamma_{abc}) \\ &= -2(\mathbf{m} \odot \mathbf{X})\mathbf{m} + \mathbf{X} = \mathbf{X}_\perp - \mathbf{X}_\parallel. \end{aligned} \tag{107}$$

Eq. (107) clearly shows the geometrical meaning of reflection. By (106), we learn reflection transformation belongs to $\text{Pin}_{p,q}$ group (**Figure 3**).

The **rotation transformation** $\mathbf{R} \in \text{Spin}_{p,q}$,

$$\mathbf{X}' = \mathbf{R}\mathbf{X}\mathbf{R}^{-1}. \tag{108}$$

The group elements of elementary transformation in Λ^2 are given by [22, 36]:

$$\left(\cosh\frac{v_{ab}}{2} + \gamma_{ab}\sinh\frac{v_{ab}}{2}\right)^{-1} = \left(\cosh\frac{v_{ab}}{2} - \gamma_{ab}\sinh\frac{v_{ab}}{2}\right), v_{ab} \in \mathbb{R}, \tag{109}$$

$$\left(\cos\frac{\theta_{ab}}{2} + \gamma_{ab}\sin\frac{\theta_{ab}}{2}\right)^{-1} = \left(\cos\frac{\theta_{ab}}{2} - \gamma_{ab}\sin\frac{\theta_{ab}}{2}\right), \theta_{ab} \in [-\pi, \pi). \tag{110}$$

The total transformation can be expressed as multiplication of elementary transformations as follows:

$$\mathbf{R} = \prod_{\{\eta_{aa}\eta_{bb}=-1\}} \left(\cosh\frac{v_{ab}}{2} + \gamma_{ab}\sinh\frac{v_{ab}}{2}\right) \prod_{\{\eta_{aa}\eta_{bb}=1\}} \left(\cos\frac{\theta_{ab}}{2} + \gamma_{ab}\sin\frac{\theta_{ab}}{2}\right). \tag{111}$$

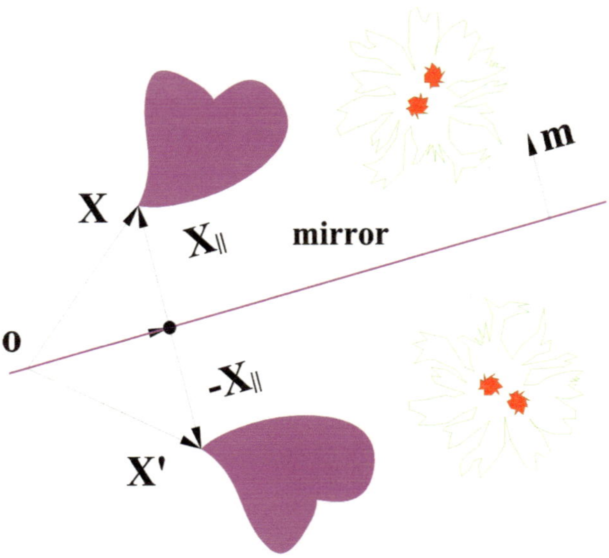

Figure 3.
Reflection transformation $\mathbf{X}' = \mathbf{X}_\perp - \mathbf{X}_\parallel$.

(111) has $\frac{1}{2}(n-1)n$ generating elements like $SO(n)$. In (111), we have commutative relation as follows:

$$\left[\cosh\frac{v_{ab}}{2}+\gamma_{ab}\sinh\frac{v_{ab}}{2},\ \cos\frac{\theta_{cd}}{2}+\gamma_{cd}\sin\frac{\theta_{cd}}{2}\right]=2\sinh\frac{v_{ab}}{2}\sin\frac{\theta_{cd}}{2}\gamma_{ab}\odot\gamma_{cd}, \tag{112}$$

$$\left[\cos\frac{\theta_{ab}}{2}+\gamma_{ab}\sin\frac{\theta_{ab}}{2},\ \cos\frac{\theta_{cd}}{2}+\gamma_{cd}\sin\frac{\theta_{cd}}{2}\right]=2\sin\frac{\theta_{ab}}{2}\sin\frac{\theta_{cd}}{2}\gamma_{ab}\odot\gamma_{cd}, \tag{113}$$

in which

$$\gamma_{ab}\odot\gamma_{cd}=\eta_{bc}\gamma_{ad}-\eta_{ac}\gamma_{bd}+\eta_{ad}\gamma_{bc}-\eta_{bd}\gamma_{ac}\in\Lambda^2. \tag{114}$$

If $a\neq b\neq c\neq d$, the right hand terms vanish, and then two elementary transformations commute with each other.

R forms a Lie Group of $\frac{1}{2}(n-1)n$ paraments. In the case $C\ell_{n,0}$ or $C\ell_{0,n}$, **R** is compact group isomorphic to $SO(n)$. Otherwise, **R** is noncompact one similar to Lorentz transformation. The infinitesimal generators of the corresponding Lie group is γ_{ab}, and the Lie algebra is given by:

$$\mathcal{R}=\varepsilon^{ab}\gamma_{ab},\quad [\gamma_{ab},\gamma_{cd}]=2\gamma_{ab}\odot\gamma_{cd}\in\Lambda^2,\quad \forall\varepsilon^{ab}\in\mathbb{R}. \tag{115}$$

Thus, $\Lambda^2(\mathbb{M}^{p,q})$ is just the Lie algebra of proper Lorentz transformation of the space-time $\mathbb{M}^{p,q}$.

5. Application in classical geometry

Suppose the basic space of projective geometry is n-dimensional Euclidean space π (see **Figure 4**), and the basis is $\{\gamma_a|a=1,2,\cdots,n\}$. The coordinate of point $\mathbf{x}$ is

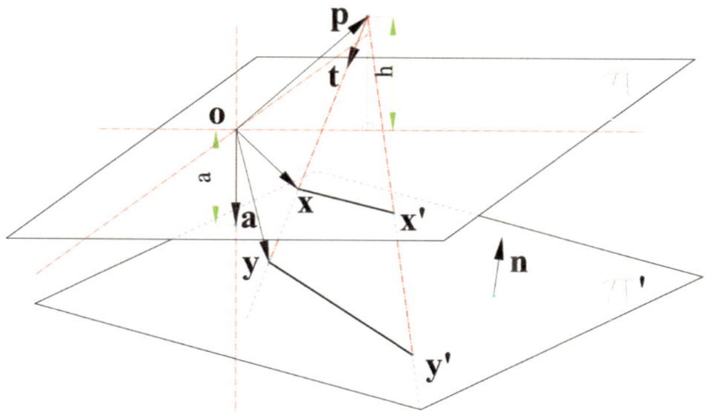

Figure 4.
Diagram of parameter setting for projective geometry.

given by $\mathbf{x} = \gamma_a x^a$. The projective polar is P, and its height from the basic space π is h. The total projective space is $n+1$ dimensional, and an auxiliary basis $\gamma_{n+1} = \gamma_p$ is introduced. The coordinate of the polar P is $\mathbf{p} = \gamma_\mu p^\mu$. In this section, we use Greek characters for $n+1$ indices. Assume the unit directional vector of the projective ray is $\mathbf{t} = \gamma_\mu t^\mu$, the unit normal vector of the image space π' is $\mathbf{n} = \gamma_\mu n^\mu$, coordinate in π' is $\mathbf{y} = \gamma_\mu y^\mu$, and the intercept of π' with the $n+1$ coordinate axis is a. Then, we have:

$$(\mathbf{y} - \mathbf{a}) \odot \mathbf{n} = 0, \quad \text{or} \quad \mathbf{y} \odot \mathbf{n} = n_\mu y^\mu = a n_p. \tag{116}$$

The equation of projective ray is given by:

$$\mathbf{s} = \mathbf{p} + \lambda \mathbf{t}, \tag{117}$$

where λ is parameter coordinate of the line. In the basic space π, we have $s^{n+1} = 0$ and $\lambda = -h/t^p$, so the coordinate of the line in π reads

$$\mathbf{x} = \mathbf{p} - \frac{h}{t^p}\mathbf{t}. \tag{118}$$

Let $\mathbf{s} = \mathbf{y}$ and substitute (117) into (116) we get image equation as follows:

$$\mathbf{y} = \mathbf{p} + \frac{a n_p - \mathbf{p} \odot \mathbf{n}}{\mathbf{t} \odot \mathbf{n}}\mathbf{t}, \qquad \lambda = \frac{a n_p - \mathbf{p} \odot \mathbf{n}}{\mathbf{t} \odot \mathbf{n}}. \tag{119}$$

In the above equation $\mathbf{t} \odot \mathbf{n} \neq 0$, which means $\mathbf{t}$ cannot be perpendicular to $\mathbf{n}$; otherwise, the projection cannot be realized. Eliminating coordinate $\mathbf{t}$ in (118) and (119), we find the projective transformation $\mathbf{y} \leftrightarrow \mathbf{x}$ is nonlinear. In (119), only the parameters $(a, \mathbf{n})$ are related to image space π'; so, all geometric variables independent of two parameters $(a, \mathbf{n})$ are projective invariants. In what follows we prove the fundamental theorems of projective geometry by Clifford algebra.

Theorem 10 *For 4 different points $\{y_1, y_2, y_3, y_4\}$ on a straight line L, the following cross ratio is a projective invariant*

$$(12; 34) \equiv \frac{|y_1 - y_3|}{|y_2 - y_3|} \cdot \frac{|y_2 - y_4|}{|y_1 - y_4|}. \tag{120}$$

Proof Substituting (119) into (120) we get

$$(12;34)=\frac{|(\mathbf{t}_3\odot\mathbf{n})\mathbf{t}_1-(\mathbf{t}_1\odot\mathbf{n})\mathbf{t}_3|}{|(\mathbf{t}_3\odot\mathbf{n})t_2-(\mathbf{t}_2\odot\mathbf{n})\mathbf{t}_3|}\cdot\frac{|(\mathbf{t}_4\odot\mathbf{n})\mathbf{t}_2-(\mathbf{t}_2\odot\mathbf{n})\mathbf{t}_4|}{|(\mathbf{t}_4\odot\mathbf{n})\mathbf{t}_1-(\mathbf{t}_1\odot\mathbf{n})\mathbf{t}_4|}. \tag{121}$$

By (19) and (20), we get

$$(\mathbf{t}_b\odot\mathbf{n})\mathbf{t}_a-(\mathbf{t}_a\odot n)\mathbf{t}_b=(\mathbf{t}_a\wedge\mathbf{t}_b)\odot\mathbf{n}=\pm|\mathbf{t}_a\wedge\mathbf{t}_b|\mathbf{m}\odot\mathbf{n}, \tag{122}$$

where $\mathbf{m}$ is the unit normal vector of the plane spanned by $(\mathbf{t}_a,\mathbf{t}_b)$, which is independent of the image space π'. Substituting it into (121), we get

$$(12;34)=\frac{|\mathbf{t}_3\wedge\mathbf{t}_1|}{|\mathbf{t}_3\wedge\mathbf{t}_2|}\cdot\frac{|\mathbf{t}_4\wedge\mathbf{t}_2|}{|\mathbf{t}_4\wedge\mathbf{t}_1|}. \tag{123}$$

(123) is independent of $(a,\mathbf{n})$; so, it is a projective invariant. Likewise, $(13;24)$ and $(14;23)$ are also projective invariants. The proof is finished.

Now we examine affine transformation. In this case, the polar P at infinity and the directional vector $\mathbf{t}$ of rays becomes constant vector. The equation of rays is given by $\mathbf{y}=\mathbf{x}+\lambda\mathbf{t}$. Substituting it into (116), we get the coordinate transformation from basic space π to image space π',

$$\mathbf{y}=\mathbf{x}+\frac{an_p-\mathbf{n}\odot\mathbf{x}}{\mathbf{t}\odot\mathbf{n}}\mathrm{t},\qquad \lambda=\frac{an_p-\mathbf{n}\odot\mathbf{x}}{\mathbf{t}\odot\mathbf{n}}. \tag{124}$$

Since $\mathbf{t}$ and $\mathbf{n}$ are constant vectors for all rays, the affine transformation $\mathbf{y}\leftrightarrow\mathbf{x}$ is linear. A variable independent of $(a,\mathbf{n})$ is an affine invariant.

Theorem 11 *Assume $\{x_1,x_2,x_3\}$ are 3 points on a straight line L in basic space π, and $\{y_1,y_2,y_3\}$ are respectively their projective images on line L' in π'. Then the simple ratio*

$$(12,13)\equiv\frac{|y_2-y_1|}{|y_3-y_1|} \tag{125}$$

is an affine invariant.

Proof By equation of transformation (124) we get

$$\mathbf{y}_k=\mathbf{x}_k+\frac{an_p-\mathbf{n}\odot\mathbf{x}_k}{\mathbf{t}\odot\mathbf{n}}\mathbf{t}. \tag{126}$$

In (126), only the parameters $(a,\mathbf{n})$ are related to image space π'. Substituting (126) into (125), we have:

$$(12,13)=\frac{|(\mathbf{t}\odot\mathbf{n})(\mathbf{x}_2-\mathbf{x}_1)-\mathbf{n}\odot(\mathbf{x}_2-\mathbf{x}_1)\mathbf{t}|}{|(\mathbf{t}\odot\mathbf{n})(\mathbf{x}_3-\mathbf{x}_1)-\mathbf{n}\odot(\mathbf{x}_3-\mathbf{x}_1)\mathbf{t}|}=\frac{|((\mathbf{x}_2-\mathbf{x}_1)\wedge\mathbf{t})\odot\mathbf{n}|}{|((\mathbf{x}_3-\mathbf{x}_1)\wedge\mathbf{t})\odot\mathbf{n}|}. \tag{127}$$

Denote the unit directional vector of line L by $\mathbf{k}$, then we have

$$\mathbf{x}_2-\mathbf{x}_1=\pm|\mathbf{x}_2-\mathbf{x}_1|\mathbf{k},\qquad \mathbf{x}_3-\mathbf{x}_1=\pm|\mathbf{x}_3-\mathbf{x}_1|\mathbf{k}. \tag{128}$$

Substituting them into (127) we get:

$$(12,13)=\frac{|\mathbf{y}_2-\mathbf{y}_1|}{|\mathbf{y}_3-\mathbf{y}_1|}=\frac{|\mathbf{x}_2-\mathbf{x}_1|}{|\mathbf{x}_3-\mathbf{x}_1|}. \tag{129}$$

This proves the simple ratio $(12, 13)$ is an affine invariant. Likewise, we can prove $(12, 23)$ and $(13, 23)$ are also affine invariants. The proof is finished.

The treatment of image information by computer requires concise and general algebraic representation for geometric modeling as well as fast and robust algebraic algorithm for geometric calculation. Conformal geometry algebra was introduced in this context. By establishing unified covariant algebra representation of classical geometry, the efficient calculation of invariant algebra is realized [13–15]. It provides a unified and concise homogeneous algebraic framework for classical geometry and algorithms, which can thus be used for complicated symbolic geometric calculations. This technology is currently widely applied in high-tech fields such as computer graphics, vision calculation, geometric design, and robots.

The algebraic representation of a geometric object is homogeneous, which means that any two algebraic expressions representing this object differ by only one nonzero factor and any such algebraic expressions with different nonzero multiple represent the same geometric object. The embedding space provided by conformal geometric algebra for n dimensional Euclidean space is $n+2$ dimensional Minkowski space. Since the orthonormal transformation group of the embedding space is exactly double coverage of the conformal transformation group of the Euclidean space, this model is also called the conformal model. The following is a brief introduction to the basic concepts and representation for geometric objects of conformal geometric algebra. The materials mainly come from literature [13].

In conformal geometry algebra, an additional Minkowski plane $\mathbb{M}^{1,1}$ is attached to n dimensional Euclidean space $\mathbb{R}^n$, $\mathbb{M}^{1,1}$ has an orthonormal basis $\{e_+, e_-\}$, which has the following properties:

$$e_+^2 = 1, \qquad e_-^2 = -1, \qquad e_+ \odot e_- = 0. \tag{130}$$

In practical application, $\{e_+, e_-\}$ is replaced by null basis $\{e_0, e\}$

$$e_0 = \frac{1}{2}(e_- - e_+), \qquad e = e_- + e_+. \tag{131}$$

They satisfy

$$e_0^2 = e^2 = 0, \qquad e \odot e_0 = -1. \tag{132}$$

A unit pseudo-scalar E for $\mathbb{M}^{1,1}$ is defined by:

$$E = e \wedge e_0 = e_+ \wedge e_- = e_+ e_-. \tag{133}$$

In conformal geometric algebra, we work with $\mathbb{M}^{n+1,1} = \mathbb{R}^n \oplus \mathbb{M}^{1,1}$.
Define the **horosphere** of $\mathbb{R}^n$ by:

$$\mathcal{N}_e^n = \{x \in \mathbb{M}^{n+1,1} | x^2 = 0, x \odot e = -1\}. \tag{134}$$

$\mathcal{N}_e^n$ is a homogeneous model of $\mathbb{R}^n$. The powerful applications of conformal geometry come from this model. By calculation, for $\forall \mathbf{x} \in \mathbb{R}^n$ we have:

$$x = \mathbf{x} + \frac{1}{2}\mathbf{x}^2 e + e_0, \tag{135}$$

which is a bijective mapping $\mathbf{x} \in \mathbb{R}^n \leftrightarrow x \in \mathcal{N}_e^n$, we have $\mathcal{N}_e^n \cong \mathbb{R}^n$. x is referred to as the **homogeneous point** of $\mathbf{x}$. Clearly, $0 \in \mathbb{R}^n \leftrightarrow e_0 \in \mathcal{N}_e^n$ and $\infty \in \mathbb{R}^n \leftrightarrow e \in \mathcal{N}_e^n$ are in homogeneous coordinate.

Now we examine how conformal geometric algebra represents geometric objects. For a line passing through points **a** and **b**, we have

$$e \wedge a \wedge b = e\mathbf{a} \wedge \mathbf{b} + (\mathbf{b} - \mathbf{a})E. \tag{136}$$

Since $\mathbf{a} \wedge \mathbf{b} = \mathbf{a} \wedge (\mathbf{b} - \mathbf{a})$ is the moment for a line through point **a** with tangent $\mathbf{a} - \mathbf{b}$, $e \wedge \mathbf{a} \wedge \mathbf{b}$ characterizes the line completely.

Again by using (135) and (136), we get

$$e \wedge a \wedge b \wedge c = e\mathbf{a} \wedge \mathbf{b} \wedge \mathbf{c} + (\mathbf{b} - \mathbf{a}) \wedge (\mathbf{c} - \mathbf{a})E. \tag{137}$$

We recognize $\mathbf{a} \wedge \mathbf{b} \wedge \mathbf{c}$ as the moment of a plane with tangent $(\mathbf{b} - \mathbf{a}) \wedge (\mathbf{c} - \mathbf{a})$. Thus, $e \wedge a \wedge b \wedge c$ represents a plane through points $\{\mathbf{a}, \mathbf{b}, \mathbf{c}\}$, or, more specifically, the triangle (2-simplex) with these points as vertices.

For a sphere with radius ρ and center $\mathbf{p} \in \mathbb{R}^n$, we have $(\mathbf{x} - \mathbf{p})^2 = \rho^2$. By (135), the equation in terms of homogeneous points becomes

$$x \odot p = -\frac{1}{2}\rho^2. \tag{138}$$

Using $x \odot e = -1$, we get:

$$x \odot s = 0, \qquad s = p - \frac{1}{2}\rho^2 e = \mathbf{p} + e_0 + \frac{1}{2}(\mathbf{p}^2 - \rho^2)e, \tag{139}$$

where

$$s^2 = \rho^2, \qquad e \odot s = -1. \tag{140}$$

From these properties, the form (139) and center **p** can be recovered. Therefore, every sphere in $\mathbb{R}^n$ is completely characterized by a unique vector $s \in \mathbb{M}^{n+1,1}$. According to (140), s lies outside the null cone. Analysis shows that every such vector determines a sphere.

6. Discussion and conclusion

The examples given above are only applications of Clifford algebra in geometry, but we have seen the power of Clifford algebra in solving geometrical problems. In fact, Clifford algebra is more widely used in physics. Why does Clifford algebra work so well? As have been seen from the above examples, the power of Clifford algebra comes from the following features:

1. In the geometry of flat space, the basic concepts are only length, angle, area, and volume, which are already implicitly included in the definition of Clifford algebra. So, Clifford algebra summarizes these contents of classical geometry and algebraize them all. By introducing the concepts of inner, exterior, and direct products of vector, Clifford algebra summarizes the operations of scalars, vectors, and tensors and then can represent all the physical variables in classical physics, because only these variables are included in classical physics.

DOI: http://dx.doi.org/10.5772/intechopen.93444

2. By localizing the basis or frame of space-time, Clifford algebra is naturally suitable for the tangent space in a manifold. If the differential ∂_μ and connection operator $\eth_\mu \gamma_\nu$ are introduced, Clifford algebra can be used for the whole manifold, so it contains Riemann geometry. Furthermore, Clifford algebra can express all contents of classical physics, including physical variables, differential equations, and algebraic operations. Clifford algebra transforms complicated theories and relations into a unified and standard calculus with no more or less contents, and all representations are neat and elegant [23, 36].

3. If the above contents seem to be very natural, Clifford algebra still has another unusual advantage, that is, it includes the theory of spinor. So, Clifford algebra also contains quantum theory and spinor connection. These things are far beyond the human intuition and have some surprising properties.

4. There are many reasons to make Clifford algebra become a unified and efficient language and tool for mathematics, physics, and engineering, such as Clifford algebra generalizes real number, complex number, quaternion, and vector algebra; Clifford algebra is isomorphic to matrix algebra; the derivative operator $\gamma^\mu \nabla_\mu$ contains grad, div, curl, etc. However, the most important feature of Clifford algebra should be taking the physical variable and the basis as one entity, such as $\mathbf{g} = g_{\mu\nu} \gamma^\mu \otimes \gamma^\nu$ and $\mathcal{T} = \mathcal{T}_{\mu\nu\omega} \gamma^{\mu\nu\omega}$. In this representation, the basis is an operator without ambiguity. Clifford algebra calculus is an arithmetic-like operation which can be well understood by everyone.

   "But, if geometric algebra is so good, why is it not more widely used?" As Hestenes replied in [11]: "Its time will come!" The published geometric algebra literature is more than sufficient to support instruction with geometric algebra at intermediate and advanced levels in physics, mathematics, engineering, and computer science. Though few faculty are conversant with geometric algebra now, most could easily learn what they need while teaching. At the introductory level, geometric algebra textbooks and teacher training will be necessary before geometric algebra can be widely taught in the schools. There is steady progress in this direction, but funding is needed to accelerate it. Malcolm Gladwell has discussed social conditions for a "tipping point" when the spread of an idea suddenly goes viral. Place your bets now on a Tipping Point for Geometric Algebra!

Acknowledgements

I would like to thank Dr. Min Lei for her kind invitation and help. The discussion on torsion is completed under the inspiration and guidance of Prof. James M. Nester. The content of conformal geometric algebra is added according to the suggestion of Dr. Isiah Zaplana. The chapter has been improved according to the comments of a referee.

Author details

Ying-Qiu Gu
School of Mathematical Science, Fudan University, Shanghai, China

*Address all correspondence to: yqgu@fudan.edu.cn

References

[1] Hamilton W. On quaternions, or on a new system of imaginaries in algebra. Philosophical Magazine. 1844

[2] Grassmann H. Die Lineale Ausdehnungslehre, ein neuer Zweig der Mathematik [The Theory of Linear Extension, a New Branch of Mathematics]. O. Wigand, 1844, New York: Cambridge University Press; 2013

[3] Clifford W. Application of Grassmann's extensive algebra. American Journal of Mathematics. 1878; **1**:350-358

[4] Gibbs W. The Scientific Papers of Willard Gibbs. Vol. 3. London: Longmas, Green and Company; 1906

[5] Dirac P. The quantum theory of the electron. Proceedings of the Royal Society of London. 1928;**A117**:610-624

[6] Cartan E. The Theory of Spinors. Cambridge, MA: The M.T.I. Press; 1966

[7] Chevalley C. The Algebraic Theory of Spinors and Clifford Algebras. Berlin: Springer; 1996

[8] Riesz M. In: Bolinder EF, Lounesto P, editors. Clifford Numbers and Spinors. Netherlands: Springer; 1993

[9] Hestenes D. Space Time Algebra. New York: Gordon and Breach; 1966

[10] Hestenes D. Oersted medal lecture 2002: Reforming the mathematical language of physics. American Journal of Physics. 2003;**71**:104-121

[11] Hestenes D. The genesis of geometric algebra: A personal retrospective. Advances in Applied Clifford Algebras. 2017;**27**(1):351-379

[12] Sobczyk G. Clifford geometric algebras in multilinear algebra and non-Euclidean geometries. In: Byrnes J, editor. Computational Noncommutative Algebra and Applications. Netherlands: Springer; 2004

[13] Li HB, Hestenes D, Rockwood A. Generalized homogeneous coordinates for computational geometry. In: Sommer G, editor. Geometric Computing with Clifford Algebras. Heidelberg: Springer-Verlag; 2001. pp. 27-60

[14] Doran C, Lasenby A. Geometric Algebra for Physicists. Cambridge: Cambridge University Press; 2003

[15] Lasenby A, Lasenby J, Wareham R. A Covariant Approach to Geometry Using Geometric Algebra, Technical Report. Cambridge, UK: University of Cambridge Department of Engineering; 2004

[16] Doran CJL, Lasenby AN, Gull SF, Somaroo S, Challinor A. Spacetime algebra and electron physics. Advances in Electronics and Electron Physics. 1996;**95**:272-383

[17] Sommer G, editor. Applications of Geometric Algebra in Engineering. 1st ed. Springer Verlag; 1999

[18] Lounesto P. Clifford Algebras and Spinors. Cambridge: Cambridge University Press; 2001

[19] Shirokov DS. Clifford algebras and their applications to Lie groups and spinors. In: Mladenov IM, Yoshioka A, editors. Proceedings of the Nineteenth International Conference on Geometry, Integrability and Quantization. Sofia: Avangard Prima; 2018. pp. 11-53. arXiv: 1709.06608

[20] Gu YQ. A Note on the Representation of Clifford Algebra. Preprints; 2020. p. 2020020466. [Preprint]. DOI: 10.20944/preprints202002.0466.v1

[21] Pavsic M. Clifford algebra, geometry and physics. NATO Science Series II. 2003;**95**:165-174. arXiv:gr-qc/0210060

[22] Lachiéze-Rey M. Spin and Clifford algebras, an introduction. Advances in Applied Clifford Algebras. 2009;**19**(3–4):687-720. arXiv:1007.2481

[23] Gu YQ. Space-time geometry and some applications of Clifford algebra in physics. Advances in Applied Clifford Algebras. 2018;**28**(4):79

[24] Hehl FW, von der Heyde P, Kerlick GD, Nester JM. General relativity with spin and torsion: Foundation and prospects. Reviews of Modern Physics. 1976;**48**:393-416

[25] Bergmann PG. Two-component Spinors in general relativity. Physics Review. 1957;**107**(**2**):624-629

[26] Gu YQ. The Simplification of Spinor Connection and Classical Approximation. [Preprint] arXiv:gr-qc/0610001

[27] Nester JM. Special orthonormal frames. Journal of Mathematical Physics. 1992;**33**:910

[28] Gu YQ. The series solution to the metric of stationary vacuum with axisymmetry. Chinese Physics B. 2008; **19**(3):90-100

[29] Blackett PMS. The magnetic field of massive rotating bodies. Nature. 1947; **159**:658-666

[30] Dolginov A. Electromagnetic field created by rotation of celestial bodies. Journal of Modern Physics. 2016;7(16): 2418-2425

[31] Gu YQ. A new explanation for the origin of magnetic field of celestial bodies. Natural Science (in Chinese). 2019;7(6):464-470

[32] Gu YQ. A canonical form for relativistic dynamic equation. Advances in Applied Clifford Algebras. 1997;**V7** (1):13-24

[33] Garling DJH. Clifford Algebras: An Introduction. Cambridge: Cambridge University Press; 2011

[34] Lawson HB, Michelsohn ML. Spin Geometry. Princeton: Princeton University Press; 1989

[35] Gunn CG. Doing Euclidean plane geometry using projective geometric algebra. Advances in Applied Clifford Algebras. 2017;**27**:1203-1232

[36] Gu YQ. Clifford algebra, Lorentz transformation and unified field theory. Advances in Applied Clifford Algebras. 2018;**28**(2):37

Chapter 5

Quasiconformal Reflections across Polygonal Lines

Samuel L. Krushkal

Abstract

An important open problem in geometric complex analysis is to establish algorithms for explicit determination of the basic curvelinear and analytic functionals intrinsically connected with conformal and quasiconformal maps, such as their Teichmüller and Grunsky norms, Fredholm eigenvalues and the quasireflection coefficient. This has not been solved even for convex polygons. This case has intrinsic interest in view of the connection of polygons with the geometry of the universal Teichmüller space and approximation theory. This survey extends our previous survey of 2005 and presents the new approaches and recent essential progress in this field of geometric complex analysis, having various important applications. Another new topic concerns quasireflections across finite collections of quasiintervals.

Keywords: Grunsky inequalities, univalent function, Beltrami coefficient, quasiconformal reflection, universal Teichmüller space, Fredholm eigenvalues, convex polygon

1. Quasiconformal reflections: general theory

1.1 Quasireflections and quasicurves

The classical Brouwer-Kerekjarto theorem ([1, 2], see also [3]) says that every periodic homeomorphism of the sphere S^2 is topologically equivalent to a rotation, or to a product of a rotation and a reflection across a diametral plane. The first case corresponds to orientation-preserving homeomorphisms (and then E consists of two points), the second one is orientation reversing, and then either the fixed point set E is empty (which is excluded in our situation) or it is a topological circle.

We are concerned with homeomorphisms reversing orientation. Such homeomorphisms of order 2 are topological involutions of S^2 with $f \circ f = \mathrm{id}$ and are called topological reflections.

We shall consider here **quasiconformal reflections** or quasireflections on the Riemann sphere $\hat{\mathbb{C}} = \mathbb{C} \cup \{\infty\} = S^2$, that is, the orientation reversing quasiconformal automorphisms of order 2 (involutions) of the sphere with $f \circ f = \mathrm{id}$. The topological circles admitting such reflections are called **quasicircles**. Such circles are locally *quasiintervals*, that is, the images of straight line segments under quasiconformal maps of the sphere S^2. Any quasireflection preserves pointwise fixed a quasicircle $L \subset \hat{\mathbb{C}}$ interchanging its inner and outer domains.

Under **quasiconformal map** $w(z)$ of a domain $D \subset \hat{\mathbb{C}}$, we understand an orientation-preserving generalized solution of the Beltrami equation (uniformly elliptic system of the first order)

$$\frac{\partial w}{\partial \bar{z}} = \mu(z) \frac{\partial w}{\partial z}, \quad z \in D,$$

where

$$\frac{\partial}{\partial z} = \frac{1}{2}\left(\frac{\partial}{\partial x} - i\frac{\partial}{\partial y}\right), \quad \frac{\partial}{\partial \bar{z}} = \frac{1}{2}\left(\frac{\partial}{\partial x} - i\frac{\partial}{\partial y}\right)$$

are the distributional partial derivatives, μ is a given function from $L_\infty(D)$ with $\|\mu\|_\infty < 1$, called the **Beltrami coefficient** (or complex dilatation) of the map w, and the quantity $k(w) = \|\mu\|_\infty$ is the (quasiconformal) **dilatation** of this map. There are some equivalent analytic and geometric definitions of such maps.

Quasiconformality preserves (up to bounded perturbations) the main intrinsic properties of conformal maps (see, e.g., [4–6]).

Qualitatively, any quasicircle L is characterized, due to [7], by uniform boundedness of the cross-ratios for all ordered quadruples (z_1, z_2, z_3, z_4) of the distinct points on L; namely,

$$\frac{\overline{z_1 z_2}}{\overline{z_1 z_3}} \frac{\overline{z_3 z_4}}{\overline{z_2 z_4}} \leq C < \infty$$

for any quadruple of points z_1, z_2, z_3, z_4 on L following this order. Using a fractional linear transformation, one can send one of the points, for example, z_4, to infinity; then the above inequality assumes the form

$$\left|\frac{z_2 - z_1}{z_3 - z_1}\right| \leq C.$$

This is shown in [7] by applying the properties of quasisymmetric maps. Ahlfors has established also that if a topological circle L admits quasireflections (i.e., is a quasicircle), then there exists a differentiable quasireflection across L which is (euclidian) bilipschitz-continuous. This property is very useful in various applications. On its extension to hyperbolic M-bilipschitz reflections see [8].

Geometrically, a quasicircle is characterized by the property that, for any two points z_1, z_2 on L, the ratio of the chordal distance $|z_1 - z_2|$ to the diameters of the corresponding subarcs with these endpoints is uniformly bounded. Note also that every quasicircle has zero two-dimensional Lebesgue measure.

Other characterizations of quasicircles are given, for example, in [9–11]. We will not touch here the extension of this theory to higher dimensions.

Quasireflections across more general sets $E \subset \hat{\mathbb{C}}$ also appear in certain questions and are of independent interest. Those sets admitting quasireflections are called **quasiconformal mirrors**.

One defines for each mirror E its **reflection coefficient**

$$q_E = \inf k(f) = \inf \quad \|\partial_z f / \partial_{\bar{z}} f\|_\infty \tag{1}$$

and **quasiconformal dilatation**

$$Q_E = (1 + q_E)/(1 - q_E) \geq 1;$$

the infimum in (1) is taken over all quasireflections across E, provided those exist, and is attained by some quasireflection f_0.

When $E = L$ is a quasicircle, the corresponding quantity

$$k_E = \inf \{k(f_*) : f_*(S^1) = E\} \tag{2}$$

and the reflection coefficient q_E can be estimated in terms of one another; moreover, due to [4, 12], we have

$$Q_E = K_E := \left(\frac{1 + k_E}{1 - k_E}\right)^2. \tag{3}$$

The infimum in (2) is taken over all orientation-preserving quasiconformal automorhisms f_* carrying the unit circle onto L, and $k(f_*) = \|\partial_{\bar{z}} f_* / \partial_z f_*\|_\infty$.

Theorem 1. *For any set $E \subset \hat{\mathbb{C}}$ which admits quasireflections, there is a quasicircle $L \supset E$ with the same reflection coefficient; therefore,*

$$Q_E = \min \{Q_L : \quad L \supset E \quad \text{quasicircle}\}. \tag{4}$$

The proof of this important theorem was given for finite sets $E = \{z_1, \ldots, z_n\}$ by Kühnau in [13], using Teichmüller's theorem on extremal quasiconformal maps applied to the homotopy classes of homeomorphisms of the punctured spheres, and extended to arbitrary sets $E \subset hC$ by the author in [14].

Theorem 1 yields, in particular, that similar to (3) for any set $E \subset \hat{\mathbb{C}}$, its quasiconformal dilatation satsfies

$$Q_E = (1 + k_E)^2/(1 - k_E)^2,$$

where $k_E = \inf \|\partial_{\bar{z}} f / \partial_z f\|_\infty$ over all quasicircles $L \supset E$ and all orientation-preserving quasiconformal homeomorphisms $f : \hat{\mathbb{C}} \to \hat{\mathbb{C}}$ with $f(\hat{\mathbb{R}}) = L$.

This theorem implies various quantitative consequences. A new its application will be given in the last section.

We point out that the conformal symmetry on the extended complex plane is strictly rigid and reduces to reflection $z \mapsto \bar{z}$ within conjugation by transformations $g \in PSL(2, \mathbb{C})$. The quasiconformal symmetry avoids such rigidity and is possible over quasicircles. Theorem 1 shows that, in fact, this case is the most general one, since for any set $E \subset \hat{\mathbb{C}}$ we have $Q_E = \infty$, unless E is a subset of a quasicircle with the same reflection coefficient.

Let us mention also that a somewhat different construction of quasiconformal reflections across Jordan curves has been provided in [15]; it relies on the conformally natural extension of homeomorphisms of the circle introduced by Douady and Earle [16].

The quasireflection coefficients of curves are closely connected with intrinsic functionals of conformal and quasiconformal maps such as their Teichmüller and Grunsky norms and the first Fredholm eigenvalue, which imply a deep quantitative characterization of the features of these maps.

One of the main problem here, important also in applications of geometric complex analysis, is to establish the algorithms for explicit determination of these quantities for individual quasicircles or quasiintervals. This was remains open a long time even for generic quadrilaterals.

1.2 Fredholm eigenvalues

Recall that the **Fredholm eigenvalues** ρ_n of an oriented smooth closed Jordan curve L on the Riemann sphere $\hat{\mathbb{C}} = \mathbb{C} \cup \{\infty\}$ are the eigenvalues of its double-layer potential, or equivalently, of the integral equation

$$u(z) + \frac{\rho}{\pi} \int_L u(\zeta) \frac{\partial}{\partial n_\zeta} \log \frac{1}{|\zeta - z|} ds_\zeta = h(z),$$

which has has many applications (here n_ζ is the outer normal and ds_ζ is the length element at $\zeta \in L$).

The least positive eigenvalue $\rho_L = \rho_1$ plays a crucial role and is naturally connected with conformal and quasiconformal maps. It can be defined for any oriented closed Jordan curve L by

$$\frac{1}{\rho_L} = \sup \frac{|\mathcal{D}_G(u) - \mathcal{D}_{G^*}(u)|}{\mathcal{D}_G(u) + \mathcal{D}_{G^*}(u)},$$

where G and G^* are, respectively, the interior and exterior of L; $\mathcal{D}$ denotes the Dirichlet integral, and the supremum is taken over all functions u continuous on $\hat{\mathbb{C}}$ and harmonic on $G \cup G^*$. In particular, $\rho_L = \infty$ only for the circle.

An upper bound for ρ_L is given by Ahlfors' inequality [17].

$$\frac{1}{\rho_L} \leq q_L, \tag{5}$$

where q_L denotes the minimal dilatation of quasireflections across L.

In view of the invariance of all quantities in (5) under the action of the Möbius group $PSL(2, \mathbb{C})/\pm \mathbf{1}$, it suffices to consider the quasiconformal homeomorphisms of the sphere carrying S^1 onto L whose Beltrami coefficients $\mu_f(z) = \partial_{\bar{z}} f / \partial_z f$ have support in the unit disk $\mathbb{D} = \{|z| < 1\}$, and $f|\mathbb{D}^*(z) = z + b_0 + b_1 z^{-1} + \ldots$, where $\mathbb{D}^* = \hat{\mathbb{C}} \backslash \overline{\mathbb{D}}$ (or in the upper half-plane $U = \{\mathrm{Im} z > 0\}$). Then q_L is equal to the minimum $k_0(f)$ of dilatations $k(f) = \|\mu\|_\infty$ of quasiconformal extensions of the function $f^* = f|\mathbb{D}^*$ into $\mathbb{D}$.

The inequality (5) serves as a background for defining the value ρ_L, being combined with the features of Grunsky inequalities given by the classical Kühnau-Schiffer theorem. The related results can be found, for example, in surveys [12, 18, 19] and the references cited there.

In the following sections, we provide a new general approach.

1.3 The Grunsky and Milin inequalities

Let

$$\mathbb{D} = \{z : |z| < 1\}, \ \mathbb{D}^* = \left\{z \in \hat{\mathbb{C}} = \mathbb{C} \cup \{\infty\} : |z| > 1\right\}.$$

In 1939, Grunsky discovered the necessary and sufficient conditions for univalence of a holomorphic function in a finitely connected domain on the extended complex plane $\hat{\mathbb{C}}$ in terms of an infinite system of the coefficient inequalities. In particular, his theorem for the canonical disk $\mathbb{D}^*$ yields that a holomorphic function $f(z) = z + const + O(z^{-1})$ in a neighborhood of $z = \infty$ can be extended to a univalent holomorphic function on the $\mathbb{D}^*$ if and only if its Grunsky coefficients α_{mn} satisfy

$$\left| \sum_{m,n=1}^{\infty} \sqrt{mn}\ \alpha_{mn} x_m x_n \right| \le 1, \tag{6}$$

where α_{mn} are defined by

$$\log \frac{f(z) - f(\zeta)}{z - \zeta} = - \sum_{m,n=1}^{\infty} \alpha_{mn} z^{-m} \zeta^{-n}, \quad (z, \zeta) \in (\mathbb{D}^*)^2, \tag{7}$$

the sequence $\mathbf{x} = (x_n)$ runs over the unit sphere $S(l^2)$ of the Hilbert space l^2 with norm $\|\mathbf{x}\|^2 = \sum_1^{\infty} |x_n|^2$, and the principal branch of the logarithmic function is chosen (cf. [20]). The quantity

$$\varkappa(f) = \sup \left\{ \left| \sum_{m,n=1}^{\infty} \sqrt{mn} \alpha_{mn} x_m x_n \right| : \ \mathbf{x} = (x_n) \in S(l^2) \right\} \le 1 \tag{8}$$

is called the **Grunsky norm** of f.

For the functions with k-quasiconformal extensions ($k < 1$), we have instead of (8) a stronger bound

$$\left| \sum_{m,n=1}^{\infty} \sqrt{mn}\ \alpha_{mn} x_m x_n \right| \le k \quad \text{for any} \quad \mathbf{x} = (x_n) \in S(l^2), \tag{9}$$

established first in [21] (see also [18, 22]). Then

$$\varkappa(f) \le k(f), \tag{10}$$

where $k(f)$ denotes the **Teichmüller norm** of f, which is equal to the infimum of dilatations $k(w^\mu) = \|\mu\|_\infty$ of quasiconformal extensions of f to $\hat{\mathbb{C}}$. Here w^μ denotes a homeomorphic solution to the Beltrami equation $\partial_{\bar{z}} w = \mu \partial_z w$ on $\mathbb{C}$ extending f.

Note that the Grunsky (matrix) operator

$$\mathcal{G}(f) = \left(\sqrt{mn}\ \alpha_{mn}(f) \right)_{m,n=1}^{\infty}$$

acts as a linear operator $l^2 \to l^2$ contracting the norms of elements $\mathbf{x} \in l^2$; the norm of this operator equals $\varkappa(f)$ (cf. [23, 24]).

For most functions f, we have in (10) the strong inequality $\varkappa(f) < k(f)$ (moreover, the functions satisfying this inequality form a dense subset of Σ), while the functions with the equal norms play a crucial role in many applications (see [18, 22, 25–28]).

The method of Grunsky inequalities was generalized in several directions, even to bordered Riemann surfaces X with a finite number of boundary components (cf. [6, 11, 20, 29, 30]; see also [31]). In the general case, the generating function (7) must be replaced by a bilinear differential

$$-\log \frac{f(z) - f(\zeta)}{z - \zeta} - R_X(z, \zeta) = \sum_{m,n=1}^{\infty} \beta_{mn}\ \varphi_m(z) \varphi_n(\zeta) : \ X \times X \to \mathbb{C}, \tag{11}$$

where the surface kernel $R_X(z,\zeta)$ relates to the conformal map $j_\theta(z,\zeta)$ of X onto the sphere $\hat{\mathbb{C}}$ slit along arcs of logarithmic spirals inclined at the angle $\theta \in [0,\pi)$ to a ray issuing from the origin so that $j_\theta(\zeta,\zeta) = 0$ and

$$j_\theta(z) = (z - z_\theta)^{-1} + \text{const} + O(1/(z - z_\theta)) \quad \text{as} \quad z \to z_\theta = j_\theta^{-1}(\infty)$$

(in fact, only the maps j_0 and $j_{\pi/2}$ are applied). Here $\{\varphi_n\}_1^\infty$ is a canonical system of holomorphic functions on X such that (in a local parameter)

$$\varphi_n(z) = \frac{a_{n,n}}{z^n} + \frac{a_{n+1,n}}{z^{n+1}} + \dots \quad \text{with} \quad a_{n,n} > 0, \quad n = 1, 2, \dots,$$

and the derivatives (linear holomorphic differentials) φ_n' form a complete orthonormal system in $H^2(X)$.

We shall deal only with simply connected domains $X = D^* \ni \infty$ with quasiconformal boundaries (quasidisks). For any such domain, the kernel R_D vanishes identically on $D^* \times D^*$, and the expansion (11) assumes the form

$$-\log \frac{f(z) - f(\zeta)}{z - \zeta} = \sum_{m,n=1}^{\infty} \frac{\beta_{mn}}{\sqrt{mn}\ \chi(z)^m\ \chi(\zeta)^n}, \tag{12}$$

where χ denotes a conformal map of D^* onto the disk $\mathbb{D}^*$ so that $\chi(\infty) = \infty,\ \chi'(\infty) > 0$.

Each coefficient $\alpha_{mn}(f)$ in (12) is represented as a polynomial of a finite number of the initial coefficients $b_1, b_2, \dots, b_s$ of f; hence it depends holomorphically on Beltrami coefficients of quasiconformal extensions of f as well as on the Schwarzian derivatives

$$S_f(z) = \left(\frac{f''(z)}{f'(z)}\right)' - \frac{1}{2}\left(\frac{f''(z)}{f'(z)}\right)^2, \quad z \in D^*. \tag{13}$$

These derivatives range over a bounded domain in the complex Banach space $\mathbf{B}(D^*)$ of hyperbolically bounded holomorphic functions $\varphi \in \mathbb{D}^*$ with norm

$$\|\varphi\|_{\mathbf{B}} = \sup_{D^*} \lambda_{D^*}^{-2}(z)|\varphi(z)|,$$

where $\lambda_{D^*}(z)|dz|$ denotes the hyperbolic metric of D^* of Gaussian curvature -4. This domain models the **universal Teichmüller space T** with the base point $\chi'(\infty)D^*$ (in holomorphic Bers' embedding of **T**).

A theorem of Milin [29] extending the Grunsky univalence criterion for the disk $\mathbb{D}^*$ to multiply connected domains D^* states that a holomorphic function $f(z) = z + \text{const} + O(z^{-1})$ in a neighborhood of $z = \infty$ can be continued to a univalent function in the whole domain D^* if and only if the coefficients β_{mn} in (12) satisfy, similar to the classical case of the disk $\mathbb{D}^*$, the inequality

$$\left| \sum_{m,n=1}^{\infty} \beta_{mn}\ x_m x_n \right| \le 1 \tag{14}$$

for any point $\mathbf{x} = (x_n) \in S(l^2)$. We call the quantity

$$\varkappa_{D^*}(f) = \sup\left\{ \left| \sum_{m,n=1}^{\infty} \beta_{mn}\; x_m x_n \right| : \quad \mathbf{x} = (x_n) \in S(l^2) \right\}, \tag{15}$$

the **generalized Grunsky norm** of f. By (14), $\varkappa_{D^*}(f) \le 1$ for any f from the class $\Sigma(D^*)$ of univalent functions in D^* with hydrodynamical normalization

$$f(z) = z + b_0 + b_1 z^{-1} + \ldots \quad \text{near} \quad z = \infty.$$

The inequality $\varkappa_{D^*}(f) \le 1$ is necessary and sufficient for univalence of f in D^* (see [11, 20, 29]).

The norm (15) also is dominated by the Teichmüller norm $k(f)$ of this map. Similar to (10),

$$\varkappa_{D^*}(f) \le k(f) = \tanh \tau_{\mathbf{T}}(\mathbf{0}, S_F),$$

where $\tau_{\mathbf{T}}$ denotes the Teichmüller distance on the universal Teichmüller space $\mathbf{T}$ with the base point D, and for the most of univalent functions, we also have here the strict inequality.

The quasiconformal theory of generic Grunsky coefficients has been developed in [32]. This technique is a powerful tool in geometric complex analysis having fundamental applications in the Teichmüller space theory and other fields.

Note that in the case $D^* = \mathbb{D}^*$, $\beta_{mn} = \sqrt{mn}\,\alpha_{mn}$; for this disk, we shall use the notations Σ and $\varkappa(f)$. We denote by S the canonical class of univalent functions $F(z) = z + a_2 z^2 + \ldots$ in the unit disk $\mathbb{D}$.

The Grunsky norm of univalent functions $F \in S$ is defined similar to (5), (6); so any such $F(z)$ and its inversion $f(z) = 1/F(1/z)$ univalent in D^* have the same Grunsky coefficients α_{mn}. Technically it is more convenient to deal with functions univalent in $\mathbb{D}^*$.

1.4 Extremal quasiconformality

A crucial point here is that the Teichmüller norm on Σ is intrinsically connected with **integrable holomorphic quadratic differentials** $\psi(z)dz^2$ on the complementary domain

$$D = \hat{\mathbb{C}} \setminus \overline{D^*}$$

(the elements of the subspace $A_1(D)$ of $L_1(D)$ formed by holomorphic functions), while the Grunsky norm naturally relates to the **abelian** structure determined by the set of quadratic differentials

$$A_1^2(D) = \{\psi \in A_1(D) : \quad \psi = \omega^2\}$$

having only zeros of even order on D.

We describe the general intrinsic features. Let L be a quasicircle passing through the points $0, 1, \infty$, which is the common boundary of two domains D and D^*. Let L be an oriented quasiconformal Jordan curve (quasicircle) on the Riemann sphere $\hat{\mathbb{C}}$ with the interior and exterior domains D and D^*. Denote by $\lambda_D(z)|dz|$ the hyperbolic metric of D of Gaussian curvature -4 and by $\delta_D(z) = dist(z, \partial D)$ the Euclidean distance from the point $z \in D$ to the boundary. Then

$$\frac{1}{4} \le \lambda_D(z)\delta_D(z) \le 1,$$

where the right-hand inequality follows from the Schwarz lemma and the left from Koebe's $\frac{1}{4}$ theorem.

Consider the unit ball of Beltrami coefficients supported on D,

$$\text{Belt}(D)_1 = \{\mu \in L_\infty(\mathbb{C}) : \ \mu|D^* = 0 \quad \|\mu\|_\infty < 1\}$$

and take the corresponding quasiconformal automorphisms $w^\mu(z)$ of the sphere $\hat{\mathbb{C}}$ satisfying on $\mathbb{C}$ the Beltrami equation $\partial_{\bar{z}} w = \mu \partial_z w$ preserving the points $0, 1, \infty$ fixed. Recall that $k(w) = \|\mu_w\|_\infty$ is the dilatation of the map w.

Take the equivalence classes $[\mu]$ and $[w^\mu]$ letting the coefficients μ_1 and μ_2 from $Belt(D^*)_1$ be equivalent if the corresponding maps w^{μ_1} and w^{μ_2} coincide on L (and hence on $\overline{D}$). These classes are in one-to-one correspondence with the Schwarzians S_{w^μ} on D^*, which fill a bounded domain in the space $\mathbf{B}_2(D^*)$ modeling the universal Teichmüller space $\mathbf{T} = \mathbf{T}(D^*)$ with the base point D^*. The quotient map

$$\phi_{\mathbf{T}} : \text{Belt}(D)_1 \to \mathbf{T}, \quad \phi_{\mathbf{T}}(\mu) = S_{w^\mu}$$

is holomorphic (as the map from $L_\infty(D)$ to $\mathbf{B}_2(D)$). Its intrinsic *Teichmüller metric* is defined by

$$\tau_{\mathbf{T}}(\phi_{\mathbf{T}}(\mu), \phi_{\mathbf{T}}(\nu)) = \frac{1}{2} \inf\left\{ \log K\left(w^{\mu_*} \circ (w^{\nu_*})^{-1}\right) : \mu_* \in \phi_{\mathbf{T}}(\mu), \nu_* \in \phi_{\mathbf{T}}(\nu)\right\},$$

It is the integral form of the infinitesimal Finsler metric

$$F_{\mathbf{T}}(\phi_{\mathbf{T}}(\mu), \phi'_{\mathbf{T}}(\mu)\nu) = \inf\left\{ \|\nu_* / \left(1 - |\mu|^2\right)\|_\infty : \ \phi'_{\mathbf{T}}(\mu)\nu_* = \phi'_{\mathbf{T}}(\mu)\nu \right\}$$

on the tangent bundle $\mathcal{T}\mathbf{T}$ of $\mathbf{T}$, which is locally Lipschitzian.

The Grunsky coefficients give rise to another Finsler structure $F(\mathbf{x}, v)$ on the bundle $\mathcal{T}\mathbf{T}$. It is dominated by the canonical Finsler structure $F_{\mathbf{T}}(\mathbf{x}, v)$ and allows one to reconstruct the Grunsky norm along the geodesic Teichmüller disks in $\mathbf{T}$ (see [33]).

We call the Beltrami coefficient $\mu \in \text{Belt}(D^*)_1$ **extremal** (in its class) if

$$\|\mu\|_\infty = \inf\{\|\nu\|_\infty : \phi_{\mathbf{T}}(\nu) = \phi_{\mathbf{T}}(\mu)\}$$

and call μ **infinitesimally extremal** if

$$\|\mu\|_\infty = \inf\left\{\|\nu\|_\infty : \nu \in L_\infty(D^*), \quad \phi'_{\mathbf{T}}(\mathbf{0})\nu = \phi'_{\mathbf{T}}(\mathbf{0})\mu\right\}.$$

Any infinitesimally extremal Beltrami coefficient μ is globally extremal (and vice versa), and by the basic Hamilton-Krushkal-Reich-Strebel theorem the extremality of μ is equivalent to the equality

$$\|\mu\|_\infty = \inf\{| < \mu, \psi >_{D^*} | : \psi \in A(D) : \|\psi\| = 1\}$$

where $A(D)$ is the space of the integrable holomorphic quadratic differentials on D (the subspace of $L_1(D)$ formed by holomorphic functions on D) and the pairing

$$\langle \mu, \psi \rangle_D = \iint_D \mu(z)\psi(z)dxdy, \quad \mu \in L_\infty(D), \ \psi \in L_1(D) \quad (z = x + iy).$$

Let $w_0 := w^{\mu_0}$ be an extremal representative of its class $[w_0]$ with dilatation

$$k(w_0) = \|\mu_0\|_\infty = \inf\{k(w^\mu) : w^\mu|L = w_0|L\},$$

and assume that there exists in this class a quasiconformal map w_1 whose Beltrami coefficient μ_{A_1} satisfies the inequality $ess\,sup_{A_r}|\mu_{w_1}(z)| < k(w_0)$ in some ring domain $\mathcal{R} = D^* \setminus G$ complement to a domain $G \supset D^*$. Any such w_1 is called the **frame map** for the class $[w_0]$, and the corresponding point in the universal Teichmüller space $\mathbf{T}$ is called the **Strebel point**.

These points have the following important properties.

Theorem 2. *(i) If a class $[f]$ has a frame map, then the extremal map f_0 in this class (minimizing the dilatation $\|\mu\|_\infty$) is unique and either a conformal or a Teichmüller map with Beltrami coefficient $\mu_0 = k|\psi_0|/\psi_0$ on D, defined by an integrable holomorphic quadratic differential ψ_0 on D and a constant $k \in (0,1)$* [34].

(ii) The set of Strebel points is open and dense in $\mathbf{T}$ [35, 36].

The first assertion holds, for example, for asymptotically conformal curves L. Similar results hold also for arbitrary Riemann surfaces (cf. [36, 37]).

Recall that a Jordan curve $L \subset \mathbb{C}$ is called **asymptotically conformal** if for any pair of points $a, b \in L$,

$$\max_{z \in L} \frac{|a - z| + |z - b|}{|a - b|} \to 1 \quad \text{as} \quad |a - b| \to 0,$$

where z lies between a and b.

Such curves are quasicircles without corners and can be rather pathological (see, e.g., [38, p. 249]). In particular, all C^1-smooth curves are asymptotically conformal.

The polygonal lines are not asymptotically conformal, and the presence of angles causes non-uniqueness of extremal quasireflections.

The boundary dilatation $H(f)$ admits also a local version $H_p(f)$ involving the Beltrami coefficients supported in the neighborhoods of a boundary point $p \in \partial D$. Moreover (see, e.g., [36], Ch. 17), $H(f) = \sup_{p \in \partial D} H_p(f)$, and the points with $H_p(f) = H(f)$ are called **substantial** for f and for its equivalence class.

On the unique and non-unique extremality see, for example, [5, 34, 39–44].

The extremal quasiconformality is naturally connected with extremal quasireflections.

1.5 Complex geometry and basic Finsler metrics on universal Teichmüller space

Recall that the universal Teichmüller space $\mathbf{T}$ is the space of quasisymmetric homeomorphisms h of the unit circle $S^1 = \partial\mathbb{D}$ factorized by Möbius transformations. Its topology and real geometry are determined by the Teichmüller metric, which naturally arises from extensions of these homeomorphisms h to the unit disk. This space admits also the complex structure of a complex Banach manifold (and this is valid for all Teichmüller spaces).

One of the fundamental notions of geometric complex analysis is the invariant Kobayashi metric on hyperbolic complex manifolds, even in the infinite dimensional Banach or locally convex complex spaces.

The canonical complex Banach structure on the space $\mathbf{T}$ is defined by factorization of the ball of Beltrami coefficients

$$\text{Belt}(\mathbb{D})_1 = \{\mu \in L_\infty(\mathbb{C}) : \mu|\mathbb{D}^* = 0, \ \|\mu\| < 1\},$$

letting $\mu, \nu \in \mathrm{Belt}(\mathbb{D})_1$ be equivalent if the corresponding maps $w^{\mu}, w^{\nu} \in \Sigma^0$ coincide on S^1 (hence, on $\overline{\mathbb{D}^*}$) and passing to Schwarzian derivatives $S_{f^{\mu}}$. The defining projection $\phi_{\mathbf{T}} : \mu \to S_{w^{\mu}}$ is a holomorphic map from $L_{\infty}(\mathbb{D})$ to $\mathbf{B}$. The equivalence class of a map w^{μ} will be denoted by $[w^{\mu}]$.

An intrinsic complete metric on the space $\mathbf{T}$ is the Teichmüller metric, defined above in Section 1.4, with its infinitesimal Finsler form (structure) $F_{\mathbf{T}}(\phi_{\mathbf{T}}(\mu), \phi'_{\mathbf{T}}(\mu)\nu), \mu \in \mathrm{Belt}(\mathbb{D})_1; \nu, \nu_* \in L_{\infty}(\mathbb{C})$.

The space $\mathbf{T}$ as a complex Banach manifold also has invariant metrics. Two of these (the largest and the smallest metrics) are of special interest. They are called the Kobayashi and the Carathéodory metrics, respectively, and are defined as follows.

The **Kobayashi metric** $d_{\mathbf{T}}$ on $\mathbf{T}$ is the largest pseudometric d on $\mathbf{T}$ does not get increased by holomorphic maps $h : \mathbb{D} \to \mathbf{T}$ so that for any two points $\psi_1, \ \psi_2 \in \mathbf{T}$, we have

$$d_{\mathbf{T}}(\psi_1, \psi_2) \leq \inf \{d_{\mathbb{D}}(0,t) : h(0) = \psi_1, \ h(t) = \psi_2\},$$

where $d_{\mathbb{D}}$ is the **hyperbolic Poincaré metric** on $\mathbb{D}$ of Gaussian curvature -4, with the differential form

$$ds = \lambda_{\mathrm{hyp}}(z)|dz| := |dz| / \left(1 - |z|^2\right).$$

The **Carathéodory** distance between ψ_1 and ψ_2 in $\mathbf{T}$ is

$$c_{\mathbf{T}}(\psi_1, \psi_2) = \sup d_{\mathbb{D}}(h(\psi_1), h(\psi_2)),$$

where the supremum is taken over all holomorphic maps $h : \ \mathbb{D} \to \mathbf{T}$.

The corresponding differential (infinitesimal) forms of the Kobayashi and Carathéodory metrics are defined for the points (ψ, v) of the tangent bundle $\mathcal{T}(\mathbf{T})$, respectively, by

$$\begin{aligned} \mathcal{K}_{\mathbf{T}}(\psi, v) &= \inf \{1/r : \ r > 0, \ h \in \mathrm{Hol}(\mathbb{D}_r, \mathbf{T}), h(0) = \psi, dh(0) = v\}, \\ \mathcal{C}_{\mathbf{T}}(\psi, v) &= \sup\{|df(\psi)v| : \ f \in \mathrm{Hol}(\mathbf{T}, \mathbb{D}), f(\psi) = 0\}, \end{aligned}$$

where $\mathrm{Hol}(X, Y)$ denotes the collection of holomorphic maps of a complex manifold X into Y and $\mathbb{D}_r$ is the disk $\{|z| < r\}$.

The Schwarz lemma implies that the Carathéodory metric is dominated by the Kobayashi metric (and similarly for their infinitesimal forms). We shall use here mostly the Kobayashi metric.

Due to the fundamental Gardiner-Royden theorem, the Kobayashi metric on any Teichmüller spaces is equal to its Teichmüller metric (see [15, 36, 40, 45]).

For the the universal Teichmüller space $\mathbf{T}$, we have the following strengthened version of this theorem for universal Teichmüller space given in [46].

Theorem 3. *The Teichmüller metric $\tau_{\mathbf{T}}(\psi_1, \psi_2)$ of either of the spaces $\mathbf{T}$ or $\mathbf{T}(\mathbb{D}^*)$ is plurisubharmonic separately in each of its arguments; hence, the pluricomplex Green function of $\mathbf{T}$ equals*

$$g_{\mathbf{T}}(\psi_1, \psi_2) = \log \tanh \tau_{\mathbf{T}}(\psi_1, \psi_2) = \log k(\psi_1, \psi_2),$$

where k is the norm of extremal Beltrami coefficient defining the distance between the points ψ_1, ψ_2 in $\mathbf{T}$ (and similar for the space $\mathbf{T}(\mathbb{D}^*)$).

The differential (infinitesimal) Kobayashi metric $\mathcal{K}_{\mathbf{T}}(\psi, v)$ on the tangent bundle $\mathcal{T}(\mathbf{T})$ of $\mathbf{T}$ is logarithmically plurisubharmonic in $\psi \in \mathbf{T}$, equals the infinitesimal Finsler *form* $F_{\mathbf{T}}(\psi, v)$ *of metric* $\tau_{\mathbf{T}}$ *and has constant holomorphic sectional curvature* $\kappa_{\mathcal{K}}(\psi, v) = -4$ *on the tangent bundle* $\mathcal{T}(\mathbf{T})$.

In other words, the Teichmüller-Kobayashi metric is the largest invariant plurisubharmonic metric on $\mathbf{T}$.

The **generalized Gaussian curvature** κ_λ of an upper semicontinuous Finsler metric $ds = \lambda(t)|dt|$ in a domain $\Omega \subset \mathbb{C}$ is defined by

$$\kappa_\lambda(t) = -\frac{\mathbb{D} \log \lambda(t)}{\lambda(t)^2},$$

where $\mathbb{D}$ is the **generalized Laplacian**

$$\mathbb{D}\lambda(t) = 4 \liminf_{r \to 0} \frac{1}{r^2} \left\{ \frac{1}{2\pi} \int_0^{2\pi} \lambda\left(t + re^{i\theta}\right) d\theta - \lambda(t) \right\}$$

(provided that $-\infty \le \lambda(t) < \infty$). Similar to C^2 functions, for which $\mathbb{D}$ coincides with the usual Laplacian, one obtains that λ is subharmonic on Ω if and only if $\mathbb{D}\lambda(t) \ge 0$; hence, at the points t_0 of local maximuma of λ with $\lambda(t_0) > -\infty$, we have $\mathbb{D}\lambda(t_0) \le 0$.

The **sectional holomorphic curvature** of a Finsler metric on a complex Banach manifold X is defined in a similar way as the supremum of the curvatures over appropriate collections of holomorphic maps from the disk into X for a given tangent direction in the image.

The holomorphic curvature of the Kobayashi metric $\mathcal{K}(x, v)$ of any complete hyperbolic manifold X satisfies $\kappa_{\mathcal{K}_X} \ge -4$ at all points (x, v) of the tangent bundle $\mathcal{T}(X)$ of X, and for the Carathéodory metric $\mathcal{C}_X$ we have $\kappa_{\mathcal{C}}(x, v) \le -4$.

Finally, the **pluricomplex Green function** of a domain X on a complex Banach space manifold E is defined as $g_X(x, y) = \sup u_y(x) \quad (x, y \in X)$, where supremum is taken over all plurisubharmonic functions $u_y(x): \ X \to [-\infty, 0)$ satisfying $u_y(x) = \log\|x - y\| + O(1)$ in a neighborhood of the pole y. Here $\|\cdot\|$ is the norm on X and the remainder term $O(1)$ is bounded from above. If X is hyperbolic and its Kobayashi metric d_X is logarithmically plurisubharmonic, then $g_X(x, y) = \log \tanh \ d_X(x, y)$, which yields the representation of $g_{\mathbf{T}}$ in Theorem 3.

For details and general properties of invariant metrics, we refer to [47, 48] (see also [18, 49]).

Theorem 3 has various applications in geometric function theory and in complex geometry Teichmüller spaces. Its proof involves the technique of the Grunsky coefficient inequalities.

Plurisubharmonicity of a function $u(x)$ on a domain D in a Banach space X means that $u(x)$ is upper continuous in D and its restriction to the intersection of D with any complex line L is subharmonic.

A deep Zhuravlev's theorem implies that the intersection of the universal Teichmüller space $\mathbf{T}$ with every complex line is a union of simply connected planar (moreover, this holds for any Teichmüller space); see ([50], pp. 75–82, [51]).

1.6 The Grunsky-Milin inequalities revised

Denote by $\Sigma^0(D^*)$ the subclass of $\Sigma(D^*)$ formed by univalent $\hat{\mathbb{C}}$-holomorphic functions in D^* with expansions $f(z) = z + b_0 + b_1 z^{-1} + \ldots$ near $z = \infty$ admitting quasiconformal extensions to $\hat{\mathbb{C}}$. It is dense in $\Sigma(D^*)$ in the weak topology of locally uniform convergence on D^*.

Each Grunsky coefficient $\alpha_{mn}(f)$ is a polynomial of a finite number of the initial coefficients $b_1, b_2, \ldots, b_{m+n-1}$ of f; hence it depends holomorphically on Beltrami coefficients of extensions of f as well as on the Schwarzian derivatives $S_f \in \mathbf{B}_2(D^*)$.

Consider the set

$$A_1^2(D) = \{\psi \in A_1(D) : \psi = \omega^2\}$$

consisting of the integrable holomorphic quadratic differentials on D having only zeros of even order and put

$$\alpha_D(f) = \sup \left\{|\langle\mu_0, \psi\rangle_D| : \psi \in A_1^2, \quad \|\psi\|_{A_1(D)} = 1\right\}.$$

The following theorem from [32] completely describes the relation between the Grunsky and Teichmüller norms (more special results were obtained in [26, 52]).

Theorem 4. *For all* $f \in \Sigma^0(D^*)$,

$$\varkappa_{D^*}(f) \le k \frac{k + \alpha_D(f)}{1 + \alpha_D(f)k}, \; k = k(f),$$

and $\varkappa_{D^*}(f) < k$ unless

$$\alpha_D(f) = \|\mu_0\|_\infty, \tag{16}$$

where μ_0 is an extremal Beltrami coefficient in the equivalence class $[f]$. The last equality is equivalent to $\varkappa_{D^*}(f) = k(f)$.

If $\varkappa(f) = k(f)$ and the equivalence class of f (the collection of maps equal to f on $S^1 = \partial D^*$) is a Strebel point, then the extremal μ_0 in this class is necessarily of the form

$$\mu_0 = \|\mu_0\|_\infty |\psi_0| / \psi_0 \quad \text{with} \quad \psi_0 \in A_1^2(D). \tag{17}$$

Note that geometrically (16) means the equality of the Carathéodory and Teichmüller distances on the geodesic disk $\{\phi_{\mathbf{T}}(t\mu_0/\|\mu_0\|) : t \in \mathbb{D}\}$ in the universal Teichmüller space $\mathbf{T}$ and that the mentioned above the strict inequality $\varkappa(f) < k(f)$ is valid on the (open) dense subset of Σ^0 in both strong and weak topologies (i.e., in the Teichmüller distance and in locally uniform convergence on D^*).

An important property of the Grunsky coefficients $\alpha_{mn}(f) = \alpha_{mn}(S_F)$ is that these coefficients are holomorphic functions of the Schwarzians $\varphi = S_f$ on the universal Teichmüller space $\mathbf{T}$. Therefore, for every $f \in \Sigma^0$ and each $\mathbf{x} = (x_n) \in S(l^2)$, the series

$$h_{\mathbf{x}}(\varphi) = \sum_{m,n=1}^{\infty} \sqrt{mn}\, \alpha_{mn}(\varphi) x_m x_n \tag{18}$$

defines a holomorphic map of the space $\mathbf{T}$ into the unit disk $\mathbb{D}$, and $\varkappa_{D^*}(F) = \sup_{\mathbf{x}} |h_{\mathbf{x}}(S_F)|$.

The convergence and holomorphy of the series (18) simply follow from the inequalities

$$\left|\sum_{m=j}^{M} \sum_{n=l}^{N} \sqrt{mn}\, \alpha_{mn} x_m x_n\right|^2 \le \sum_{m=j}^{M} |x_m|^2 \sum_{n=l}^{N} |x_n|^2$$

(for any finite M, N), which, in turn, are a consequence of the classical area theorem (see, e.g., [11], p. 61; [29], p. 193).

Using Parseval's equality, one obtains that the elements of the distinguished set $A_1^2(\mathbb{D})$ are represented in the form

$$\psi(z) = \frac{1}{\pi} \sum_{m+n=2}^{\infty} \sqrt{mn}\ x_m x_n z^{m+n-2} \tag{19}$$

with $\mathbf{x} = (x_n) \in l^2$ so that $\|\mathbf{x}\|_{l^2} = \|\psi\|_{A_1}$ (see [52]). This result extends to arbitrary domains D with quasiconformal boundaries but the proof is much more complicated (see [22]).

1.7 The first Fredholm eigenvalue and Grunsky norm

One of the basic tools in quantitative estimating the Freholm eigenvalues ρ_L of quasicircles is given by the classical Kühnau-Schiffer theorem mentioned above. This theorem states that *the value ρ_L is reciprocal to the Grunsky norm $\varkappa(f)$ of the Riemann mapping function of the exterior domain of L* (see. [27, 53]).

Another important tool is the following Kühnau's jump inequality [12]:

If a closed curve $L \subset \hat{\mathbb{C}}$ contains two analytic arcs with the interior intersection angle $\pi\alpha'$, then

$$\frac{1}{\rho_L} \geq |1 - |\alpha'||. \tag{20}$$

This implies the lower estimate for q_L and $1/\rho_L$. By approximation, this inequality extends to smooth arcs.

One of the standard ways of establishing the reflection coefficients q_L (respectively, the Fredholm eigenvalues ρ_L) consists of verifying wether the equality in (5) or the equality $\varkappa(f^*) = k_0(f^*)$ hold for a given curve L (cf. [12, 28, 52, 54, 55]).

This was an open problem a long time even for the rectangles stated by R. Kühnau, after it was established only [12, 55] that the answer is in affirmative for the square and for close rectangles $\mathcal{R}$ whose moduli $m(\mathcal{R})$ vary in the interval $1 \leq m(\mathcal{R}) < 1.037$; moreover, in this case $q_L = 1/\rho_L = 1/2$. The method exploited relied on an explicit construction of an extremal reflection. The complete answer was given in [33].

The relation between the basic curvelinear functionals intrinsically connected with conformal and quasiconformal maps is described in Kühau's paper [56].

1.8 Holomorphic motions

Let E be a subset of $\hat{\mathbb{C}}$ containing at least three points.

A **holomorphic motion** of E is a function $f : E \times \mathbb{D} \to \hat{\mathbb{C}}$ such that:

a. for every fixed $z \in E$, the function $t \mapsto f(z,t) : E \times \mathbb{D} \to \hat{\mathbb{C}}$ is holomorphic in $\mathbb{D}$;

b. for every fixed $t \in \mathbb{D}$, the map $f(z,t) = f_t(z) : E \to \hat{\mathbb{C}}$ is injective;

c. $f(z,0) = z$ for all $z \in E$.

The remarkable lambda-lemma of Mañé, Sad, and Sullivan [57] yields that such holomorphic dependence on the time parameter provides quasiconformality of f in the space parameter z. Moreover: *(i) f extends to a holomorphic motion of the closure $\overline{E}$ of E;*

(ii) each $f_t(z) = f(t,z) : \overline{E} \to \hat{\mathbb{C}}$ is quasiconformal; (iii) f is jointly continuous in (z,t).

Quasiconformality here means, in the general case, the boundedness of the distortion of the circles centered at the points $z \in E$ or of the cross-ratios of the ordered quadruples of points of E.

The Slodkowski lifting theorem ([58], see also [59–61]) solves the problem of Sullivan and Thurston on the extension of holomorphic motions from any set to a whole sphere:

Extended lambda-lemma: *Any holomorphic motion $f : E \times \mathbb{D} \to \hat{\mathbb{C}}$ can be extended to a holomorphic motion $\tilde{f} : \hat{\mathbb{C}} \times \mathbb{D} \to \hat{\mathbb{C}}$, with $\tilde{f}|E \times \mathbb{D} = f$.*

The corresponding Beltrami differentials $\mu_{\tilde{f}_t}(z) = \partial_{\bar{z}}\tilde{f}(z,t)/\partial_z\tilde{f}(z,t)$ are holomorphic in t via elements of $L_\infty(\mathbb{C})$, and Schwarz's lemma yields

$$\|\mu_{\tilde{f}_t}\|_\infty \le |t|,$$

or equivalently, the maximal dilatations $K\left(\tilde{f}_t\right) \le (1+|t|)/(1-|t|)$. This bound cannot be improved in the general case.

Holomorphic motions have been important in the study of dynamical systems, Kleinian groups, holomorphic families of conformal maps and of Riemann surfaces as well as in many other fields (see, e.g., [40, 57, 59, 60, 62–68], and the references there).

There is an intrinsic connection between holomorphic motions and Teichmüller spaces, first mentioned by Bers and Royden in [69]. McMullen and Sullivan introduced in [65] the Teichmüller spaces for arbitrary holomorphic dynamical systems, and this approach is now one of the basic in complex dynamics [70].

Topics discussed in this section were studied in classic works [71–85] as well as other references.

2. Unbounded convex polygons

2.1 Main theorem

The inequalities (5), (20) have served a long time as the main tool for establishing the exact or approximate values of the Fredholm value ρ_L and allowed to establish it only for some special collections of curves and arcs.

In this section, we present, following [33, 86], a new method that enables us to solve the indicated problems for large classes of convex domains and of their fractional linear images. This method involves in an essential way the complex geometry of the universal Teichmüller space $\mathbf{T}$ and the Finsler metrics on holomorphic disks in $\mathbf{T}$ as well as the properties of holomorphic motions on such disks.

It is based on the following general theorem for unbounded convex domains giving implies an explicit representation of the main associated curvelinear and analytic functionals invariants by geometric characteristics of these domains solving the problem for unbounded convex domains completely.

Theorem 5. *For every unbounded convex domain $D \subset \mathbb{C}$ with piecewise $C^{1+\delta}$-smooth boundary L $(\delta > 0)$ (and all its fractional linear images), we have the equalities*

$$q_L = 1/\rho_L = \varkappa(f) = \varkappa(f^*) = k_0(f) = k_0(f^*) = 1 - |\alpha|, \qquad (21)$$

where f and f^* denote the appropriately normalized conformal maps $\mathbb{D} \to D$ and $\mathbb{D}^* \to D^* = \hat{\mathbb{C}} \setminus \overline{D}$, respectively, $k_0(f)$ and $k_0(f^*)$ are the minimal dilatations of their quasiconformal extensions to $\hat{\mathbb{C}}$; $\varkappa(f)$ and $\varkappa(f^*)$ stand for their Grunsky norms, and $\pi|\alpha|$ is the opening of the least interior angle between the boundary arcs $L_j \subset L$. Here $0 < \alpha < 1$ if the corresponding vertex is finite and $-1 < \alpha < 0$ for the angle at the vertex at infinity.

The same is true also for the unbounded concave domains (the complements of convex ones) which do not contain ∞*; for those one must replace the last term by* $|\beta| - 1$*, where* $\pi|\beta|$ *is the opening of the largest interior angle of* D*.*

The proof of Theorem 5 is outlined in [33, 64]. In the next section we provide an extension of this important theorem to nonconvex polygons giving the detailed proof.

The equalities of type (21) were known earlier only for special closed curves (see [12, 19, 26, 55]), for example, for polygons bounded by circular arcs with a common inner tangent circle. The proof of Theorem 4 involves a completely different approach; it relies on the properties of holomorphic motions.

Let us mention also that the geometric assumptions of Theorem 4 are applied in the proof in an essential way. Its assertion extends neither to the arbitrary unbounded nonconvex or nonconcave domains nor to the arbitrary bounded convex domains.

This theorem has various important consequences. It distinguishes a broad class of domains, whose geometric properties provide the explicit values of intrinsic conformal and quasiconformal characteristics of these domains.

2.2 Examples

1. Let L be a closed unbounded curve with the convex interior, which is $C^{1+\delta}$ smooth at all finite points and has at infinity the asymptotes approaching the interior angle $\pi\alpha < 0$. For any such curve, Theorem 4 yields the equalities

$$q_L = 1/\rho_L = 1 - |\alpha|. \tag{22}$$

2. More generally, assume that L also has a finite angle point z_0 with the angle opening $\pi\alpha_0$. Then, similar to (22),

$$q_L = 1/\rho_L = \max(1-|\alpha_0|, 1-|\alpha_\infty|).$$

Simultaneously this quantity gives the exact value of the reflection coefficient for any convex curvelinear lune bounded by two smooth arcs with the common endpoints a, b, because any such lune is a Moebius image of the exterior domain for the above curve L.

Other quantitative examples illustrating Theorem 5 are presented in [64].

3. Extension to unbounded non-convex polygons

3.1 An open question

An open question is to establish the extent in which Theorem 5 can be prolonged to arbitrary unbounded polygons

Our goal is to show that this is possible for unbounded rectilinear polygons for which the extent of deviation from convexity is sufficiently small.

This extension essentially increases the collections of individual polygonal curves and arcs with explicitly established Fredholm eigenvalues and reflection coefficients.

3.2 Main theorem

Let P_n be a rectilinear polygon with the finite vertices $A_1, A_2, \ldots, A_{n-1}$ and with vertex $A_\infty = \infty$, and let the interior angle at the vertex A_j be equal to $\pi\alpha_j$ and at A_∞ be equal to $\pi\alpha_\infty$, where $\alpha_\infty < 0$ and all $a_j \neq 1$, so that $\alpha_1 + \ldots + \alpha_{n-1} + \alpha_\infty = 2$. Let f_n be the conformal map of the upper half-plane $U = \{z : \operatorname{Im} z > 0\}$ onto P_n, which without loss of generality, can be normalized by $f_n(z) = z - i + O(z - i)$ as $z \to i$ (assuming that P_n contains the origin $w = 0$).

An important geometric characteristic of polynomials is the quantity

$$|1 - |\alpha|| = \max \ \{|1-|\alpha_1||, \ldots, |1-|\alpha_{n-1}||, \ 1-|\alpha_\infty|||\}; \tag{23}$$

it valuates the local boundary quasiconformal dilatation of P_n.

Using this quantity, we first prove that an assertion similar to Theorem 4 fails for the generic rectilinear polygons.

Theorem 6. *There exist rectilinear polygons P_n whose conformal mapping functions f_n satisfy*

$$\varkappa(f_n) = k(f_n) > |1 - |a||, \tag{24}$$

where a is defined via (23).

Proof. We shall use the rectangles P_4; in this case all $\alpha_j = 1/2$. It is known that the mapping function f_4 of any rectangle has equal Grunsky and Teichmüller norms,

$$\varkappa(f_4) = k(f_4)$$

(see [12, 55, 87]).

Using the Moebius map $\sigma : z \mapsto 1/z$, we transform the rectangle into a (nonconvex) circular quadrilateral $\sigma(P_4)$ with angles $\pi/2$ and mutually orthogonal edges so that two unbounded edges from these are rectilinear and two bounded are circular, and note that for sufficiently long rectangles must be

$$k\left(\hat{f}_4\right) = \varkappa\left(\hat{f}_4\right) > 1/2, \tag{25}$$

where $\hat{f}_4$ denotes the conformal map $\mathbb{D} \to \sigma(P_4)$.

Indeed, as was established by Kühnau [12], the quadrilaterals with the side ratios (conformal module) greater than 3.31 have the reflection coefficient $q_{\partial P_4} > 1/2$ (the last inequality follows also from the fact that the long rectangles give in the limit a half-strip with two unbounded parallel sides. Such a domain is not a quasidisk, so its reflection coefficient equals 1); this proves (25).

Any circular quadrilateral $\sigma(P_4)$ satisfying (25) can be approximated by appropriate rectilinear polygons P_n. Assuming now that the equalities of type (21) or (24) are valid for all such polygons, one obtains a contradiction with (25), because both dilatation $k(f)$ and $q_{\partial P}$ are lower continuous functionals under locally uniform convergence of quasiconformal maps (i.e., $k(f) \leq \lim\ \inf k(f_n)$ as $f_n \to f$ in the indicated topology, and similarly for the reflection coefficient). This contradiction proves the theorem.

3.3 The main result

The main result of this section is

Theorem 7. [86] *Let P_n be a unbounded rectilinear polygon, neither convex nor concave, and hence contain the vertices A_j whose inner angles $\pi\alpha_j$ have openings $\pi\alpha_j$ with $1 < \alpha_j < 2$. Assume that all such α_j satisfy*

$$\alpha_j - 1 < |1 - |\alpha||, \tag{26}$$

where α is given by (23) (which means that the maximal value in (22) is attained at some vertex A_j with $0 < |a_j| < 1$).

For any such polygon, taking appropriate Moebius map $\sigma : \mathbb{D} \to U$, we have the equalities

$$\varkappa(f_n \circ \sigma) = k(f_n) = q_{\partial P_n} = 1/\rho_{\partial P_n} = |1 - |\alpha||. \tag{27}$$

Proof. Let P_n be an unbounded rectilinear polygon. Its conformal mapping function $f_n : U \to P_n$ fixing the infinite point and with $f_n(i) = 0$ is represented by the Schwarz-Christoffel integral

$$f_n(\zeta) = d_1 \int_0^z (\xi - a_1)^{\alpha_1 - 1} \dots (\xi - a_{n-1})^{\alpha_{n-1} - 1} d\xi + d_0, \tag{28}$$

where all $a_j = f_*^{-1}(A_j) \in \mathbb{R}$ and d_0, d_1 are the corresponding complex constants. The logarithmic derivative $b_f = (\log f')' = f''/f'$ of this map has the form

$$b_{f_n}(z) = \sum_1^{n-1} (\alpha_j - 1)(z - a_j)^{-1}.$$

Letting $I_\alpha = \{t \in \mathbb{R} : -1/|1-|\alpha|| < t < 1/|1-|\alpha||\}$, $\mathbb{D}_\alpha = \{t \in \mathbb{C} : |t| < 1/|1-|\alpha||\}$, we construct for f_n an ambient complex isotopy (holomorpic motion)

$$w(z,t) : U \times \mathbb{D}_\alpha \to \hat{\mathbb{C}}, \tag{29}$$

(containing f_n as a fiber map), which is injective in the space coordinate z for any fixed t, holomorphic in t for a fixed z and $w(z, 0) \equiv z$.

First observe that for real $r \in I_\alpha$ the solution W_r to the equation $w''(z) = rb_{f_4}(z)w'(z)$ with the initial conditions $w_r(i) = i$, $w_r(\infty) = \infty$ satisfies

$$b_{W_r}(z) = \sum_1^{n-1} r \frac{\alpha_j - 1}{z - a_j} = \sum_1^{n} \frac{\alpha_j(r) - 1}{z - a_j},$$

where

$$\alpha_j(r) = r(\alpha_j - 1) + 1. \tag{30}$$

If the interior angles of the initial polygon P_n satisfy the assumption (26), then all the functions W_r are represented by an integral of type (27) (replacing α_j by $\alpha_j(r)$, and with suitable constants d_{0r}, d_{1r}).

Geometrically this means that the exterior angle $2\pi - \pi\alpha_j(r)$ at any finite vertex $A_j(r)$ decreases with r (but the value $\alpha_j(r) - 1$ increases if $1 < \alpha_j < 2$). Under the assumption (26), the admissible bounds for the possible values of angles ensure the univalence of this integral on U for every indicated t. This yields that every $W_r(U)$ also is a polygon with the interior angles $\pi\alpha_j(r)$ for $r \neq 0$, while $W_0(U) = U$.

Now we pass to the conformal map $g_n(\zeta) = f_n \circ \sigma_0(\zeta)$ of the unit disk $\mathbb{D}$ onto P_n, using the function $\sigma_0(\zeta) = (1+\zeta)/(1-\zeta)$. This map is represented similar to (28) by

$$g_n(\zeta) = d_1 \int_0^{\zeta} \prod_1^n (\xi - e_j)^{\alpha_j - 1} d\xi + d_0,$$

where the points e_j are the preimages of vertices $e_j = g_n^{-1}(A_j)$ on the unit circle $\{|\zeta| = 1\}$. Pick d_1 to have $g_n'(0) = 1$. For this function, we have a natural complex isotopy

$$\tilde{w}_t(\zeta) = \frac{1}{t} g_n(t\zeta) : \mathbb{D} \times \mathbb{D} \to \mathbb{C}, \tag{31}$$

with

$$b_{\tilde{w}}(\zeta) = \frac{\tilde{w}_t''(\zeta)}{\tilde{w}_t'(\zeta)} = t\frac{g''(t\zeta)}{g_n'(t\zeta)} = tb_{g_n}(t\zeta). \tag{32}$$

Following (31), we set for $t = re^{i\theta}$,

$$\tilde{w}_t(\zeta) = e^{-i\theta} W_r \circ \sigma_0(e^{i\theta}\zeta).$$

The relations (32) yield that this function also is univalent in $\mathbb{D}$.

The corresponding Schwarzians $S_{\tilde{w}_r}(\zeta) = rb'_{\tilde{w}_r}(\zeta) - r^2 b_{\tilde{w}_r}(\zeta)^2/2$ fill a real analytic line Γ in the universal Teichmüller space $\mathbf{T}$ (modeled as a bounded domain in the complex Banach space $\mathbf{B}$ of hyperbolically bounded holomorphic functions on $\mathbb{D}$). This line is located in the holomorphic disk $\tilde{\Omega} = \mathbf{b}(G) \subset \mathbf{T}$, where $\mathbf{b}$ denotes the map $t \mapsto S_{\tilde{w}_t}$ and $G \supset I_\alpha$ is a simply connected planar domain.

By Zhuravlev's theorem (see [50, 51]), this domain contains for each $r \in I_\alpha$ also the points $S_{\tilde{w}_t}$ with $|t| \leq r$ (representing the curvelinear polygons with piecewise analytic boundaries).

This generates the holomorphic motions (complex isotopies) $\tilde{w}(\zeta, t) : \mathbb{D} \times G \to \hat{\mathbb{C}}$ and $w(z,t) : U \to \hat{\mathbb{C}}$ with $w(z,1) = f_n(z)$.

The basic lambda-lemma for holomorphic motions implies that every fiber map $w_t(z)$ is the restriction to U of a quasiconformal automorphism $\hat{W}_t(z)$ of the whole sphere $\hat{\mathbb{C}}$, and the Beltrami coefficients

$$\mu(z,t) = \partial_{\bar{z}} \hat{W}_t(z) / \partial_z \hat{W}_t(z), \quad t \in \mathbb{D}_\alpha,$$

in the lower half-plane $U^* = \{z : \ \operatorname{Im} z < 0\}$ depend holomorphically on t as elements of the space $L_\infty(U^*)$.

So we have a holomorphic map $\mu(\cdot, t)$ from the disk $\mathbb{D}_\alpha$ into the unit ball of Beltrami coefficients supported on U^*,

$$\operatorname{Belt}(U^*)_1 = \{\mu \in L_\infty(\mathbb{C}) : \mu(z)|U = 0, \quad \|\mu\| < 1\},$$

and the classical Schwarz lemma implies the estimate

$$k(\hat{W}_t) = \|\mu_{\hat{W}_t}\|_\infty \le |1 - |\alpha|| |t|.$$

It follows that the extremal dilatation of the initial map $f_n(z) = \hat{W}_1(z)|U$ satisfies

$$k(f_n) \le |1 - |\alpha||.$$

Hence, also $q_{\partial P_n} \le |1 - |\alpha||$ and by the inequality (10), $\varkappa(f_n) \le |1 - |\alpha||$.
On the other hand, Kühnau's lower bound (20) implies

$$\frac{1}{\rho_{\partial P_n}} \ge |1 - |\alpha||.$$

Together with (5), this yields that the polygon P_n admits all equalities (27) completing the proof of the theorem.

3.4 Some applications

Theorem 6 widens the collections of curves with explicitly given Fredholm eigenvalues and reflection coefficients.

For example, let L be a saw-tooth quasicircle with a finite number of triangular and trapezoidal teeth joined by rectilinear segments. We assume that the angles of these teeth satisfy the condition (26). Then we have the following consequence of Theorem 7.

Corollary 1. *For any quasicircle L of the indicated form, its quasireflection coefficient* q_L *and Fredholm eigenvalue* ρ_L *are given by*

$$q_L = 1/\rho_L = |1 - |\alpha||,$$

where $|\alpha|$ *is defined similar to (22) by angles between the subintervals of L. The same is valid for images* $\gamma(L)$ *under the Moebius maps* $\gamma \in PSL(2, \mathbb{C})$.

4. Connection with complex geometry of universal Teichmüller space

4.1 Introductory remarks

Another reason why the convex polygons are interesting for quasiconformal theory is their close geometric connection with the geometry of universal Teichmüller space.

There is an interesting still unsolved completely question on shape of holomorphic embeddings of Teichmüller spaces stated in [88]:

For an arbitrary finitely or infinitely generated Fuchsian group Γ is the Bers embedding of its Teichmüller space $\mathbf{T}(\Gamma)$ starlike?

Recall that in this embedding $\mathbf{T}(\Gamma)$ is represented as a bounded domain formed by the Schwarzian derivatives S_w of holomorphic univalent functions $w(z)$ in the lower half-plane $U^* = \{z : \operatorname{Im} z < 0\}$ (or in the disk) admitting quasiconformal extensions to the Riemann sphere $\hat{\mathbb{C}} = \mathbb{C} \cup \{\infty\}$ compatible with the group Γ acting on U^*.

It was shown in [89] that universal Teichmüller space $\mathbf{T} = \mathbf{T}(\mathbf{1})$ has points that cannot be joined to a distinguished point even by curves of a considerably general form, in particular, by polygonal lines with the same finite number of rectilinear segments. The proof relies on the existence of conformally rigid domains established by Thurston in [90] (see also [91]).

This implies, in particular, that universal Teichmüller space is not starlike with respect to any of its points, and there exist points $\varphi \in \mathbf{T}$ for which the line interval $\{t\varphi : 0 < t < 1\}$ contains the points from $\mathbf{B} \setminus \mathbf{S}$, where $\mathbf{B} = \mathbf{B}(U^*)$ is the Banach space of hyperbolically bounded holomorphic functions in the half-plane U^* with norm

$$\|\varphi\|_{\mathbf{B}} = 4 \sup_{U^*} y^2 |\varphi(z)|$$

and $\mathbf{S}$ denotes the set of all Schwarzian derivatives of univalent functions on U^*. These points correspond to holomorphic functions on U^* which are only locally univalent.

Toki [92] extended the result on the nonstarlikeness of the space $\mathbf{T}$ to Teichmüller spaces of Riemann surfaces that contain hyperbolic disks of arbitrary large radius, in particular, for the spaces corresponding to Fuchsian groups of second kind. The crucial point in the proof of [92] is the same as in [89].

On the other hand, it was established in [46] that also all finite dimensional Teichmüller spaces $\mathbf{T}(\Gamma)$ of high enough dimensions are not starlike.

The nonstarlikeness causes obstructions to some problems in the Teichmüller space theory and its applications to geometric complex analysis.

The argument exploited in the proof of Theorems 4 and 5 provide much simpler constructive proof that the universal Teichmüller space is not starlike, representing explicitly the functions, which violate this property. It reveals completely different underlying geometric features.

Pick unbounded convex rectilinear polygon P_n with finite vertices $A_1, \ldots, A_{n-1}$ and $A_n = \infty$. Denote the exterior angles at A_j by $\pi\alpha_j$ so that $\pi < \alpha_j < 2\pi$, $j = 1, \ldots, n-1$. Then, similar to (22), the conformal map f_n of the lower half-plane $H^* = \{z : \operatorname{Im} z < 0\}$ onto the complementary polygon $P_n^* = \hat{\mathbb{C}} \setminus \overline{P_n}$ is represented by the Schwarz-Christoffel integral

$$f_n(z) = d_1 \int_0^z (\xi - a_1)^{\alpha_1 - 1} (\xi - a_2)^{\alpha_2 - 1} \ldots (\xi - a_{n-1})^{\alpha_{n-1} - 1} d\xi + d_0,$$

with $a_j = f_n^{-1}(A_j) \in \mathbb{R}$ and complex constants d_0, d_1; here $f_n^{-1}(\infty) = \infty$. Its Schwarzian derivative is given by

$$S_{f_n}(z) = b'_{f_n}(z) - \frac{1}{2} b^2_{f_n}(z) = \sum_1^{n-1} \frac{C_j}{(z - a_j)^2} - \sum_{j,l=1}^{n-1} \frac{C_{jl}}{(z - a_j)(z - a_l)}, \qquad (33)$$

where $b_f = f''/f'$ and

$$C_j = -(\alpha_j - 1) - (\alpha_j - 1)^2/2 < 0, \quad C_{jl} = (\alpha_j - 1)(\alpha_l - 1) > 0.$$

It defines a point of the universal Teichmüller space $\mathbf{T}$ modeled as a bounded domain in the space $\mathbf{B}(H^*)$ of hyperbolically bounded holomorphic functions on H^* with norm $\|\varphi\|_{\mathbf{B}(H^*)} = \sup_{H^*} |z - \bar{z}|^2 |\varphi(z)|$.

Denote by r_0 the positive root of the equation

$$\frac{1}{2}\left[\sum_{1}^{n-1}(\alpha_j-1)^2+\sum_{j,l=1}^{n-1}(\alpha_j-1)(\alpha_l-1)\right]r^2-\sum_{1}^{n-1}(\alpha_j-1)\ r-2=0,$$

and put $S_{f_n,t}=tb'_{f_n}-b^2_{f_n}/2,\ t>0$. Then for appropriate α_j, we have.

Theorem 8. [93] *For any convex polygon* P_n, *the Schwarzians* rS_{f_n,r_0} *define for any* $0<r<r_0$ *a univalent function* $w_r: H^* \to \mathbb{C}$ *whose harmonic Beltrami coefficient* $\nu_r(z)=-(r/2)y^2S_{f_n,r_0}(\overline{z})$ *in* H *is extremal in its equivalence class, and*

$$k(w_r)=\varkappa(w_r)=\frac{r}{2}\|S_{f_n,r_0}\|_{\mathbf{B}(H^*)}. \tag{34}$$

By the Ahlfors-Weill theorem [94], every $\varphi\in\mathbf{B}(H^*)$ with $\|\varphi\|_{\mathbf{B}(H^*)}<1/2$ is the Schwarzian derivative S_W of a univalent function W in H^*, and this function has quasiconformal extension onto the upper half-plane $H=\{z:\ Imz>0\}$ with Beltrami coefficient of the form

$$\mu_\varphi(z)=-2y^2\varphi(\overline{z}),\quad \varphi=S_f\ \ (z=x+iy\in H^*)$$

called harmonic. Theorem 7 yields that any w_r with $r<r_0$ does not admit extremal quasiconformal extensions of Teichmüller type, and in view of extremality of harmonic coefficients $\mu_{S_{w_r}}$ the Schwarzians S_{w_r} for some r between r_0 and 1 must lie outside of the space $\mathbf{T}$; so this space is not a starlike domain in $\mathbf{B}(H^*)$.

4.2 There are unbounded convex polygons P_n for which the equalities (33) are valid in the strengthened form

$$k(f_n)=\varkappa(f_n)=\frac{1}{2}\|S_{f_n}\|_{\mathbf{B}(H^*)} \tag{35}$$

for all $r\le 1$, completing the bounds (21).

We illustrate this on the case of triangles. Let P_3 be a triangle with vertices $A_1, A_2\in\mathbb{R}$ and $A_3=\infty$ and exterior angles $\alpha_1,\alpha_2,\alpha_3$. The logarithmic derivative of conformal map $f_3: H^*\to P_3^*$ has the form

$$b_{f_3}(z)=\frac{\alpha_1-1}{z-a_1}+\frac{\alpha_2-1}{z-a_2}$$

with $a_j=f_3^{-1}(A_j)\in\mathbb{R},\ j=1,2$, and similar to (34),

$$S_{f_3}(z)=\frac{C_1}{(z-a_1)^2}+\frac{C_2}{(z-a_2)^2}-\frac{C_{12}}{(z-a_1)(z-a_2)}$$

with

$$C_j=-(\alpha_j-1)-\frac{1}{2}(\alpha_j-1)^2=-\frac{\alpha_j^2+1}{2}<0,\quad j=1,2;$$

$$C_{12}=(\alpha_1-1)(\alpha_2-1)>0.$$

If the angles of P_3^* satisfy $\alpha_1, \alpha_2 < |a_3|$, where $-\pi\alpha_3$ is the angle at A_3, the arguments from [93] yield that the harmonic Beltrami coefficient $\mu_{S_{f_3}}$ satisfies (35).

Surprisingly, this construction is closely connected also with the weighted bounded rational approximation in sup norm [95, 96].

5. Quasiconformal features and fredholm eigenvalues of bounded convex polygons

5.1 Affine deformations and Grunsky norm

As it was mentioned above, there exist bounded convex domains even with analytic boundaries L whose conformal mapping functions have different Grunsky and Teichmüller norms, and therefore, $\rho_L < 1/q_L$.

The aim of this chapter is to provide the classes of bounded convex domains, especially polygons, for which these norms are equal and give explicitly the values of the associate curve functionals $k(f)$, $\varkappa(f)$, q_L, ρ_L.

One of the interesting questions is *whether the equality of Teichmüller and Grunsky norms is preserved under the affine deformations*

$$g^c(w) = c_1 w + c_2 \overline{w} + c_3$$

with $c = c_2/c_1$, $|c| < 1$ (as well as of more general maps) of quasidisks.

In the case of unbounded convex domains, this follows from Theorem 4. We establish this here for bounded domains D.

More precisely, we consider the maps g^c, which are conformal in the complementary domain $D^* = \hat{\mathbb{C}} \backslash \overline{D}$ and have in D a constant quasiconformal dilatation c, regarding such maps as the **affine deformations** and the collection of domains $g^c(D)$ as the affine class of D.

If f is a quasiconformal automorphism of $\hat{\mathbb{C}}$ conformal in $\mathbb{D}^*$ mapping the disk $\mathbb{D}$ onto a domain D, then for a fixed c the maps $g^c|D \circ f$ and $(g^c \circ f)|\mathbb{D}$ differ by a conformal map $a: \ D \to g^c(D)$ and hence have in the disk $\mathbb{D}$ the same Beltrami coefficient.

Note that the inequality $|c| < 1$ equivalent to $|c_2| < |c_1|$ follows immediately from the orientation preserving under this map and its composition with conformal map by forming the corresponding affine deformation (which arises after extension the constant Beltrami coefficient c by zero to the complementary domain).

The following theorem solves the problem positively.

Theorem 9. *For any function $f \in \Sigma^0$ with $\varkappa(f) = k(f)$ mapping the disk $\mathbb{D}^*$ onto the complement of a bounded domain (quasidisk) D and any affine deformation g^c of this domain (with $|qc| < 1$), we have the equality*

$$\varkappa(g^c \circ f) = k(g^c \circ f). \tag{36}$$

Theorems 9 essentially increases the set of quasicircles $L \subset \hat{\mathbb{C}}$ for which $\rho_L = 1/q_L$ giving simultaneously the explicit values of these curve functionals. Even for quadrilaterals, this fact was known until now only for some special types of them (for rectangles [12, 27, 28, 33] and for rectilinear or circular quadrilaterals having a common tangent circle [55]).

5.2 Scheme of the proof of Theorem 9

The proof follows the lines of Theorem 1.1 in [97] and is divided into several lemmas.

First, we establish some auxiliary results characterizing the homotopy disk of a map with $\varkappa(f) = k(f)$.

Take the generic homotopy function

$$f_t(z) = tf(z/t) = z + b_0 t + b_1 t^2 z^{-1} + b_2 t^3 z^{-2} + \ldots : \ \mathbb{D}^* \times \mathbb{D} \to \hat{\mathbb{C}}.$$

Then $S_{f_t}(z) = t^{-2} S_f(t^{-1}z)$ and this point-wise map determines a holomorphic map $\chi_f(t) = S_{f_t}(\cdot) : \mathbb{D} \to \mathbf{T}$ so that the homotopy disks $\mathbb{D}(S_f) = \chi_f(\mathbb{D})$ foliate the space $\mathbf{T}$. Note also that

$$\alpha_{mn}(f_t) = \alpha_{mn}(f) t^{m+n},$$

and if $F(z) = 1/f(1/z)$ maps the unit disk onto a convex domain, then all level lines $f(|z| = r)$ for $z \in \mathbb{D}^*$ are starlike.

Lemma 1. *If the homotopy function f_t of $f \in \Sigma^0$ satisfy $\varkappa(f_{t_0}) = k(f_{t_0})$ for some $0 < t_0 < 1$, then the equality $\varkappa(f_t) = k(f_t)$ holds for all $|t| \le t_0$ and the homotopy disk $\mathbb{D}(S_{f_t})$ has no critical points t with $0 < |t| < t_0$.*

Take the univalent extension f_1 of f to a maximal disk $\mathbb{D}_b^* = \{z \in \hat{\mathbb{C}} : |z| > b\}$, $(0 < b < 1)$ and define

$$f^*(z) = b^{-1} f_1(bz) \in \Sigma^0, \quad |z| > 1.$$

Its Beltrami coefficient in $\mathbb{D}$ is defined by holomorphic quadratic differentials $\psi \in A_1^2$ of the form (19), and we have the holomorphic map, for a fixed $\mathbf{x}^b = (x_n^b) \in l^2$,

$$h_{\mathbf{x}^b}\left(S_{f_t^*}\right) = \sum_{m,n=1}^{\infty} \sqrt{mn}\, \alpha_{mn}(f^*) x_m^b x_n^b (bt)^{m+n} \tag{37}$$

of the disk $\mathbb{D}\left(S_{f^*}\right)$ into $\mathbb{D}$. In view of our assumption on f, the series (37) is convergent in some wider disk $\{|t| < a\} (a > 1)$.

Using the map (37), we pull back the hyperbolic metric $\lambda_{\mathbb{D}}(t) = |dt|/\left(1 - |t|^2\right)$ to the disk $\mathbb{D}(S_{F_1})$ (parametrized by t) and define on this disk the conformal metric $ds = \lambda_{\tilde{h}_{\mathbf{x}}}(t)|dt|$ with

$$\lambda_{\tilde{h}_{\mathbf{x}^b}}(t) = \left(h_{\mathbf{x}^a} \circ \chi_{f_1}\right)^* \lambda_{\mathbb{D}} = \frac{|\tilde{h}'_{\mathbf{x}^b}(t)||dt|}{1 - \left|\tilde{h}_{\mathbf{x}^b}(t)\right|^2}. \tag{38}$$

of Gaussian curvature -4 at noncritical points. In fact, this is the supporting metric at $t = a$ for the upper envelope $\lambda_{\varkappa} = \sup_{\mathbf{x} \in S(l^2)} \lambda_{\tilde{h}_{\mathbf{x}^b}}(t)$ of metrics (38) followed by its upper semicontinuous regularization

$$\lambda_{\varkappa}(t) \mapsto \lambda_{\varkappa}^*(t) = \limsup_{t' \to t} \lambda_{\varkappa}(t')$$

(supporting means that $\lambda_{\tilde{h}_{\mathbf{x}^b}}(a) = \lambda_{\varkappa}(a)$ and $\lambda_{\tilde{h}_{\mathbf{x}^b}}(t) < \lambda_{\varkappa}(t)$ in a neighborhood of a).

The metric $\lambda_{\varkappa}(t)$ is logarithmically subharmonic on $\mathbb{D}$ and its **generalized Laplacian**

$$\Delta u(t) = 4\lim_{r\to 0}\inf \frac{1}{r^2}\left\{\frac{1}{2\pi}\int_0^{2\pi} u\left(t + re^{i\theta}\right)d\theta - \lambda(t)\right\}$$

satisfies

$$\Delta \log \lambda_{\varkappa} \geq 4\lambda_{\varkappa}^2$$

(while for $\lambda_{\tilde{h}_{\mathbf{x}^b}}$ we have at its noncritical points $\Delta \log \lambda_{\tilde{h}_{\mathbf{x}^b}} = 4\lambda_{\tilde{h}_{\mathbf{x}^b}}^2$).

As was mentioned above, the Grunsky coefficients define on the tangent bundle $\mathcal{T}(\mathbf{T})$ a new Finsler structure $F_{\varkappa}(\varphi, v)$ dominated by the infinitesimal Teichmüller metric $F(\varphi, v)$. This structure generates on any embedded holomorphic disk $\gamma(\mathbb{D}) \subset \mathbf{T}$ the corresponding Finsler metric $\lambda_{\gamma}(t) = F_{\varkappa}(\gamma(t), \gamma'(t))$ and reconstructs the Grunsky norm by integration along the Teichmüller disks:

Lemma 2. [97] *On any extremal Teichmüller disk $\mathbb{D}(\mu_0) = \{\phi_{\mathbf{T}}(t\mu_0) : t \in \mathbb{D}\}$ (and its isometric images in $\mathbf{T}$), we have the equality*

$$\tanh^{-1}[\varkappa(f^{r\mu_0})] = \int_0^r \lambda_{\varkappa}(t)dt.$$

Taking into account that the disk $\mathbb{D}(S_f)$ touches at the point $\varphi = S_{f_a}$ the Teichmüller disk centered at the origin of $\mathbf{T}$ and passing through this point and that the metric $\lambda_{\varkappa}$ does not depend on the tangent unit vectors whose initial points are the points of $\mathbb{D}(S_f)$, one obtains from Lemma 3 and the equality $\varkappa(f_a) = k(f_a)$ that also

$$\lambda_{\varkappa}(a) = \lambda_{\mathcal{K}}(a). \tag{39}$$

The following lemma is a needed reformulation of Theorem 3.

Lemma 3. [97] *The infinitesimal forms $\mathcal{K}_{\mathbf{T}}(\varphi, v)$ and $F_{\mathbf{T}}(\varphi, v)$ of both Kobayashi and Teichmüller metrics on the tangent bundle $\mathcal{T}(\mathbf{T})$ of $\mathbf{T}$ are continuous logarithmically plurisubharmonic in $\varphi \in \mathbf{T}$ and have constant holomorphic sectional curvature $\kappa_{\mathcal{K}}(\varphi, v) = -4$.*

We compare the metric $\lambda_{\tilde{h}_{\mathbf{x}^b}}$ with $\lambda_{\mathcal{K}}$ using Lemmas 2, 3, and Minda's maximum principle given by.

Lemma 4. [98] *If a function $u : D \to [-\infty, +\infty)$ is upper semicontinuous in a domain $D \subset \mathbb{C}$ and its (generalized) Laplacian satisfies the inequality $\Delta u(z) \geq Ku(z)$ with some positive constant K at any point $z \in D$, where $u(z) > -\infty$, and if*

$$\lim_{z\to\zeta} \sup u(z) \leq 0 \quad \textit{for all} \quad \zeta \in \partial D,$$

then either $u(z) < 0$ for all $z \in D$ or else $u(z) = 0$ for all $z \in \Omega$.

Lemma 4 and the equality (39) imply that the metrics $\lambda_{\tilde{h}_{\mathbf{x}^b}}, \lambda_{\varkappa}, \lambda_{\mathcal{K}}$ must be equal in the entire disk $\mathbb{D}(S_F)$, which yields by Lemma 2 the equality

$$\varkappa(f_r) = k(f_r) = \left|\sum_{m,n=1}^{\infty} \sqrt{mn}\,\alpha_{mn}(F_1)r^{m+n}x_m^r x_n^r\right|$$

for all $r = |t| \in (0, 1)$ (with $(x_n^r) \in S(l^2)$ depending on r) and that for any $f \in \Sigma^0$ with $\varkappa(f) = k(f)$ *its homotopy disk $\mathbb{D}(S_F)$ has only a singularity at the origin of $\mathbf{T}$.*

We may now investigate the action of affine deformations on the set of functions $f \in \Sigma^0$ with equal Grunsky and Teichmüller norms.

Lemma 5. *For any affine deformation g^c of a convex domain D with expansion $g^c(w) = w + b_0^c + b_1^c w^{-1} + \ldots$ near $w = \infty$, we have*

$$b_1^c = \frac{S_{g^c}(\infty)}{6} = \frac{1}{6} \lim_{z \to \infty} w^4 S_{g^c}(w) \neq 0,$$

and for sufficiently small $|c|$ all composite maps

$$W_{f,c}(z) = g^c \circ f(z) = z + \hat{b}_0^c + \hat{b}_1^c z^{-1} + \ldots, \quad f \in \Sigma^0,$$

also satisfy $\hat{b}_1^c \neq 0$.

Finally, we use the following important result of Kühnau [27].

Lemma 6. *For any function $f(z) = z + b_0 + b_1 z^{-1} + \ldots \in \Sigma^0$ with $b_1 \neq 0$, the extremal quasiconformal extensions of the homotopy functions f_t to $\mathbb{D}$ are defined for sufficiently small $|t| \le r_0 = r_0(f)$ $(r_0 > 0)$ by nonvanishing holomorphic quadratic differentials, and therefore, $\varkappa(f_t) = k(f_t)$.*

Using these lemmas, one establishes the equalities $\lambda_\varkappa = \lambda_\mathcal{K}$ on the disk $\mathbb{D}\left(S_{W_{f,c}}\right)$ and

$$\varkappa(W_{F,c}) = k(W_{F,c}). \tag{40}$$

The final step of the proof is to extend the last equality to all c with $|c| < 1$.

Applying again the chain rule for Beltrami coefficients μ, ν from the unit ball in $L_\infty(\mathbb{C})$,

$$w^\mu \circ w^\nu = w^\tau \quad \text{with } \tau = (\nu + \tilde{\mu})/(1 + \overline{\nu}\tilde{\mu})$$

and $\tilde{\mu}(z) = \mu(w^\nu(z))\overline{w_z^\nu}/w_z^\nu$ (so for ν fixed, τ depends holomorphically on μ in L_∞ norm) and defining the corresponding functions (37), one gets now the holomorphic functions of $c \in \mathbb{D}$. Then, constructing in a similar way the corresponding Finsler metrics

$$\lambda_{\tilde{h}_\mathbf{x}}(c) = |\tilde{h}_\mathbf{x}'(c)||dc|/\left(1 - \left|\tilde{h}_\mathbf{x}(c)\right|^2\right), \quad |c| < 1.$$

and taking their upper envelope $\lambda_\varkappa(c)$ and its upper semicontinuous regularization, one obtains a subharmonic metric of Gaussian curvature $\kappa_{\lambda_\varkappa} \le -4$ on the nonsingular disk $\{|c| < 1\}$. One can repeat for this metric all the above arguments using the already established equality (40) for small $|c|$.

5.3 Generalization

The arguments in the proof of Theorem 9 are extended almost straightforwardly to more general case:

Theorem 10. *Let $F \in \Sigma^0$ and $\varkappa(F) = k(F)$. Let h be a holomorphic map $\mathbb{D} \to \mathbf{T}$ without critical points in $\mathbb{D}$ and $h(0) = S_F$. Denote by $\mathbf{g}^c$ the univalent solution of the Schwarzian equation*

$$S_\mathbf{g} = (h(c) \circ H)(H')^2 + S_H,$$

where $H(w) = F^{-1}(w)$, on the domain $F(\mathbb{D}^*)$. Then, for any $c \in \mathbb{D}$, the composition $\mathbf{g}^c \circ F$ also satisfies $\varkappa(\mathbf{g}^c \circ F) = k(\mathbf{g}^c \circ F)$.

Note that by the lambda lemma for holomorphic motions, the map h determines a holomorphic disk in the ball of Beltrami coefficients on $F(\mathbb{D})$, which yields, together with assumptions of the theorem, that for small $|c|$,

$$\mathbf{g}^c(w) = w + b_0^c + b_1^c w^{-1} + \dots \quad \text{as } w \to \infty$$

with $b_1^c \neq 0$. This was an essential point in the proof.

5.4 Bounded polygons

The case of bounded convex polygons has an intrinsic interest, in view of the following negative fact underlying the features and contrasting Theorem 5.

Theorem 11. *There exist bounded rectilinear convex polygons P_n with sufficiently large number of sides such that*

$$\rho_{\partial P_n} < 1/q_{\partial P_n}.$$

It follows simply from Theorem 8 that if a polygon P_n, whose edges are quasiconformal arcs, satisfies $\rho_{\partial P_n} = 1/q_{\partial P_n}$ then this equality is preserved for all its affine images. In particular, this is valid for all rectilinear polygons obtained by affine maps from polygons with edges having a common tangent ellipse (which includes the regular n-gons).

Theorem 10 naturally gives rise to the question whether the property $\rho_{\partial P_n} = 1/q_{\partial P_n}$ is valid for all bounded convex polygons with sufficiently small number of sides.

In the case of triangles this immediately follows from Theorem 7 as well as from Werner's result.

Noting that the affinity preserves parallelism and moves the lines to lines, one concludes from Theorem 8 that the equality $\rho_{\partial P_4} = 1/q_{\partial P_4}$ holds in particular for quadrilaterals P_4 obtained by affine transformations from quadrilaterals that are symmetric with respect to one of diagonals and for quadrilaterals whose sides have common tangent outwardly ellipse (in particular, for all parallelograms and trapezoids). For the same reasons, it holds also for hexagons with axial symmetry having two opposite sides parallel to this axes.

In fact, Theorem 8 allows us to establish much stronger result answering the question positively for quadrilaterals.

Theorem 12. *For every rectilinear convex quadrilateral P_4, we have*

$$\varkappa(f) = k(f) = \rho_{\partial P_4} = 1/q_{\partial P_4}, \tag{41}$$

where F is the appropriately normalized conformal map of $\mathbb{D}^*$ onto P_4^*.

The proof of this theorem essentially relies on Theorem 8 and on result of [33] that the equalities (41) are valid for all rectangles, and hence for their affine transformations.

Fix such a quadrilateral $P_4^0 = A_1^0 A_2^0 A_3^0 A_4^0$ and consider the collection $\mathcal{P}^0$ of quadrilaterals $P_4 = A_1^0 A_2^0 A_3^0 A_4$ with the same first three vertices and variable A_4; the corresponding A_4 runs over a subset E of the trice punctured sphere $\hat{\mathbb{C}} \setminus \{A_1^0, A_2^0, A_3^0\}$.

The collection $\mathcal{P}^0$ contains the trapezoids, for which we have the equalities (41) by Theorem 8 (and consequently, the infinitesimal equality (39) at the corresponding points a).

Similar to the proof of Theorem 6, one obtains in the universal Teichmüller space **T** a holomorphic disk Ω extending the real analytic curve filled by the Schwarzians, which correspond to the values $t = A$ on E. On this disk, one can construct, similar to (38), the corresponding metric $\lambda_{\varkappa}$. Lemmas 4–6 again imply that this metric must coincide at all points of Ω with the dominant infinitesimal Teichmüller-Kobayashi metric λ_K of **T**. Together with Lemma 2, this provides the global equalities (41) for all points of the disk Ω (and hence for the prescribed quadrilateral P_4^0).

5.5 An open problem here is the following question of Kühnau (personal communication)

Question: *Does the reflection coefficient of a rectangle $\mathcal{R}$ be a monotone nondecreasing function of its conformal module $\mu_{\mathcal{R}}$ (the ratio of the vertical and horizontal side lengths)?*

The results of Kühnau and Werner for the rectangles $\mathcal{R}$ state that if the module $\mu(\mathcal{R})$ satisfies $1 \leq \mu(\mathcal{R}) < 1.037$, then

$$q_{\partial\mathcal{R}} = 1/\rho_{\partial\mathcal{R}} = 1/2;$$

if $\mu(\mathcal{R}) > 2.76$, then $q_{\partial\mathcal{R}} > 1/2$ (see [12, 55]).

On the other hand, the reflection coefficients of long rectangles are close to 1, because the limit half-strip is not a quasidisk.

6. Reflections across finite collections of quasiintervals

6.1 General comments

There are only a few exact estimates of the reflection coefficients of quasiconformal arcs (quasiintervals) and some their sharp upper bounds presented in [14, 99]. The most of these bounds have been obtained using the classical Bernstein-Walsh-Siciak theorem, which quantitatively connects holomorphic extension of a function defined on a compact $K \Subset \mathbb{C}^n$ with the speed of its polynomial approximation. Another approach was applied by Kühnau in [54, 100–102]. In particular, using somewhat modification of Teichmüller's Verschiebungssatz [103], he established in [102] the reflection coefficient of the set E, which consists of the interval $[-2i, 2i]$ and a separate point $t > 0$. All these results are presented in [64].

Theorems 4 and 6 open a new way in solving this problem following the lines of the first example after Theorem 4.

6.2 Reflections across the finite collections of quasiintervals

Theorems 5 and 7 open a new way in solving this problem following the lines of the first example after Theorem 5. Namely, given a finite union

$$L = \cup L_1 \cup L_2 \ldots \cup L_N$$

of smooth curvelinear quasiintervals (possibly mutually separated) such that L can be extended without adding new vertices (angular points) to a quasicircle $L_0 \supset L$ containing $z = \infty$ and bounding a convex polygon P_N that satisfies the assumptions of Theorem 4 or a polygon considered in Theorem 7, then by these theorems, *the reflection coefficient of the set L equals*

$$q_L = |1 - |a||, \tag{42}$$

where α is defined for L_0 similar to (23).

The main point here is to get a **convex** (or sufficiently close to convex, as in Theorem 7) polygon, because the initial and final arcs of components L_j can be smoothly extended and then rounded off.

Note also that adding to L a finite number of appropriately located isolated points $z_1, \ldots z_m$ does not change the reflection coefficient (42).

Additional Classification

2010 Mathematics Subject Classification: Primary: 30C55, 30C62, 30F60; **Secondary**: 31A35, 58B15

Author details

Samuel L. Krushkal[1,2]

1 Department of Mathematics, Bar-Ilan University, Ramat-Gan, Israel

2 Department of Mathematics, University of Virginia, Charlottesville, VA, USA

*Address all correspondence to: krushkal@math.biu.ac.il

References

[1] Brouwer LEF. Über die periodischen Transformationen der Kugel. Mathematische Annalen. 1919;**80**:39-41

[2] Grunsky H. Koeffizienten-bediengungen für schlicht abbildende meromorphe Funktionen. Mathematische Zeitschrift. 1939;**45**: 29-61

[3] Smith PA. Transformations of finite period. Annals of Mathematics. 1938;**39**: 127-164

[4] Ahlfors LV. Lectures on Quasiconformal Mappings. Princeton: Van Nostrand; 1966

[5] Krushkal SL. Grunsky coefficient inequalities, Carathéodory metric and extremal quasiconformal mappings. Commentarii Mathematici Helvetici. 1989;**64**:650-660

[6] Lebedev NA. The Area Principle in the Theory of Univalent Functions. Moscow: Nauka; 1975 (Russian)

[7] Ahlfors LV. Quasiconformal reflections. Acta Mathematica. 1963; **109**:291-301

[8] Gehring FW. Characteristic Properties of Quasidisks. Les Presses de l'Université de Montréal; 1982

[9] Gardiner FP, Lakic N. Quasiconformal Teichmüller Theory. Providence, RI: American Mathematical Society; 2000

[10] Lehto O, Virtanen KL. Quasikonforme Abbildungen. Berlin: Springer-Verlag; 1965

[11] Pommerenke C. Univalent Functions. Göttingen: Vandenhoeck & Ruprecht; 1975

[12] Kühnau R. Möglichst konforme Spiegelung an einer Jordankurve. Jahresbericht der Deutschen Mathematiker-Vereinigung. 1988;**90**: 90-109

[13] Kühnau R. Interpolation by extremal quasiconformal Jordan curves. Siberian Mathematical Journal. 1991;**32**: 257-264

[14] Krushkal SL. Quasiconformal mirrors. Siberian Mathematical Journal. 1999;**40**:742-753

[15] Earle CJ, Mitra S. Variation of moduli under holomorphic motions. In: In the tradition of Ahlfors and Bers (Stony Brook, NY, 1998), Contemp. Math. **256**. Providence, RI: Americal Mathematical Society; 2000. pp. 39-67

[16] Douady A, Earle CJ. Conformally natural extension of homeomorphisms of the circle. Acta Mathematica. 1986; **157**:23-48

[17] Ahlfors LV. Remarks on the Neumann-Poincaré equation. Pacific Journal of Mathematics. 1952;**2**:271-280

[18] Krushkal SL. Plurisubharmonic features of the Teichmüller metric. Publications de l'Institut Mathématique-Beograd, Nouvelle série. 2004;**75(89)**: 119-138

[19] Kühnau R. Einige neuere Entwicklungen bei quasikonformen Abbildungen. Jahresbericht der Deutschen Mathematiker-Vereinigung. 1992;**94**:141-169

[20] Grinshpan AZ. Logarithmic geometry, exponentiation, and coefficient boundsin the theory of univalent functions and nonoverlapping damains, Ch. 11. In: Kühnau R, editor. Handbook of Complex Analysis: Geometric Function Theory. Vol. 1. Amsterdam: Elsevier Science; 2002. pp. 273-332

[21] Kühnau R. Verzerrungssatze und Koeffizientenbedingungen vom Grunskyschen Typ für quasikonforme Abbildungen. Mathematische Nachrichten. 1971;**48**:77-105

[22] Krushkal SL. Complex homotopy and Grunsky operator. Complex Variables Elliptic Equations. 2014;**59**: 48-58

[23] Goluzin GM. Geometric theory of functions of complex variables. In: Transl. of Math. Monographs. Vol. 26. Providence, RI: Amer. Math. Soc.; 1969

[24] Grinshpan AZ. Remarks on the Grunsky norm and *p*th roor transformation, continued fractions and geometric function theory (CONFUN) (Tronheim, 1997). Journal of Computational and Applied Mathematics. 1999;**105**(1–2):311-315

[25] Krushkal SL, Kühnau R. Grunsky inequalities and quasiconformal extension. Israel Journal of Mathematics. 2006;**152**:49-59

[26] Kühnau R. Quasikonforme Fortsetzbarkeit, Fredholmsche Eigenwerte und Grunskysche Koeffizientenbedingungen. Annales Academiae Scientiarum Fennicae. Ser. A I. Mathematica. 1982;**7**:383-391

[27] Kühnau R. Wann sind die Grunskyschen Koeffizienten-bedingungen hinreichend für *Q*-quasikonforme Fortsetzbarkeit? Commentarii Mathematici Helvetici. 1986;**61**:290-307

[28] Kühnau R. Zur Berechnung der Fredholmschen Eigenwerte ebener Kurven. ZAMM. 1986;**66**:193-201

[29] Milin IM. Univalent functions and orthonormal systems. In: Transl. of Mathematical Monographs. Vol. 49, Transl. of Odnolistnye funktcii i normirovannie systemy. Providence, RI: American Mathematical Society; 1977

[30] Schiffer M, Spencer D. Functionals of Finite Riemann Surfaces. Princeton: Princeton University Press; 1954

[31] Gehring FW, Hag K. Reflections on reflections in quasidisks. In: Papers on Analysis: A Volume Dedicated to Olli Martio on the Occasion of his 60th Birthday, Report. Vol. 83. Univ. Jyväskylä; 2001. pp. 81-90

[32] Krushkal SL. Strengthened Grunsky and Milin inequalities. Contemporary Mathematics. 2016;**667**:159-179

[33] Krushkal SL. Quasireflections, Fredholm eigenvalues and Finsler metrics. Doklady Mathematics. 2004; **69**:221-224

[34] Strebel K. Zur Frage der Eindeutigkeit extremaler quasikonformer Abbildungen des Einheitskreises. II. Commentarii Mathematici Helvetici. 1964;**39**:77-89

[35] Lakic N. Strebel Points, Lipa's Legacy, Contemporary Mathematics **211**. Providence, RI: American Mathematical Society; 2001. pp. 417-431

[36] Gardiner FP. Teichmüller Theory and Quadratic Differentials. New York: Wiley-Interscience; 1987

[37] Earle CJ, Li Z. Isometrically embedded polydisks in infinite dimensional Teichmüller spaces. Journal of Geometric Analysis. 1999;**9**:51-71

[38] Pommerenke C. Boundary Behavior of Conformal Maps. Berlin: Springer; 1992

[39] Božin V, Lakic N, Marković V, Mateljević M. Unique extremality. Journal d'Analyse Mathématique. 1998; **75**:299-338

[40] Earle CJ, Kra I, Krushkal SL. Holomorphic motions and Teichmüller spaces. Transactions of the American

Mathematical Society. 1994;**944**: 927-948

[41] Mateljević M. Quasiconformal maps and Teichmüller theory—Extremal mappings, Overview II. Bulletin T. CXLV De l"Académie Serbe Des Sciences et Des Arts. 2013;**38**:129–172. Available from: https://www.emis.de/journals/BSANU/

[42] Reich E. On the mapping with complex dilatation $e^{i\theta}$. Annales Academiae Scientiarum Fennicae. Ser. A I. Mathematica. 1987;**12**:261-268

[43] Strebel K. Zur Frage der Eindeutigkeit extremaler quasikonformer Abbildungen des Einheitskreises. Commentarii Mathematici Helvetici. 1962;**36**:306-323

[44] Tanigawa H. Holomorphic families of geodesic discs in infinite dimensional Teichmüller spaces. Nagoya Mathematical Journal. 1992;**127**:117-128

[45] Royden HL. Automorphisms and isometries of Teichmller space. In: Advances in the Theory of Riemann Surfaces (Ann. of Math. Stud., vol. 66). Princeton: Princeton University Press; 1971. pp. 369-383

[46] Krushkal SL. Teichmüller spaces are not starlike. Annales Academiae Scientiarum Fennicae. Ser. A I. Mathematica. 1995;**20**:167-173

[47] Dineen S. The Schwarz Lemma. Oxford: Clarendon Press; 1989

[48] Kerekjarto B. Über die periodischen Transformationen der Kreisscheibe und der Kugelfläche. Mathematische Annalen. 1919;**80**:36-38

[49] Abate M, Patrizio G. Isometries of the Teichmüller metric. Ann. Scuola Super. Pisa Cl. Sci. 1998;**26**(4):437-452

[50] Krushkal SL, Kühnau R. Quasikonforme Abbildungen - neue Methoden und Anwendungen. In: Teubner-Texte zur Math. Vol. 54. Leipzig: Teubner; 1983

[51] Zhuravlev IV. Univalent Functions and Teichmüller Spaces. Novosibirsk: Institute of Mathematics; 1979. pp. 23. (Russian) preprint

[52] Krushkal SL. Quasiconformal Mappings and Riemann Surfaces. New York: Wiley; 1979

[53] Schiffer M. Fredholm eigenvalues and Grunsky matrices. Annales Polonici Mathematici. 1981;**39**:149-164

[54] Kühnau R. Möglichst konforme Spiegelung an einem Jordanbogen auf der Zahlenkugel, complex analysis. In: Hersch J, Huber A, editors. Pfluger Anniversary Volume. Basel: Birkhäuser; 1988. pp. 139-156

[55] Werner S. Spiegelungskoeffizient und Fredholmscher Eigenwert für gewisse Polygone. Annales Academiae Scientiarum Fennicae. Ser. A I. Mathematica. 1997;**22**:165-186

[56] Kühnau R. Drei Funktionale eines Quasikreises. Annales Academiae Scientiarum Fennicae. Ser. A I. Mathematica. 2000;**25**:413-415

[57] Mañé R, Sad P, Sullivan D. On the dynamics of rational maps. Annales Scientifiques de l'École Normale Supérieure. 1983;**16**:193-217

[58] Slodkowski Z. Holomorphic motions and polynomial hulls. Proceedings of the American Mathematical Society. 1991;**111**:347-355

[59] Astala K, Iwaniec T, Martin G. Elliptic Differential Equations and Quasiconformal Mappings in the Plane. Princeton, NJ: Princeton University Press; 2009

[60] Chirka EM. Holomorphic motions and uniformization of holomorphic

families of Riemann surfaces. Russian Mathematical Surveys. 2012;**67**: 1091-1165

[61] Douady A. Prolongement de mouvements holomorphes [d'après Slodkowski et autres]. Séminaire N. Bourbaki, 1993–1994. Exposé 775. pp. 1-12.

[62] Mitra S. On extensions of holomorphic motionsa survey. Geometry of Riemann Surfaces. In: London Math. Soc. Lecture Note Ser., 368. Cambridge: Cambridge University Press; 2010. pp. 283-308

[63] Mitra S, Sudeb, Shiga H. Extensions of holomorphic motions and holomorphic families of Mbius groups. Osaka Journal of Mathematics. 2010;**47**: 1167-1187

[64] Krushkal SL. Quasiconformal extensions and reflections, Ch 11. In: Kühnau R, editor. Handbook of Complex Analysis: Geometric Function Theory. Vol. 2. Amsterdam: Elsevier Science; 2005. pp. 507-553

[65] McMullen C, Sullivan D. Quasiconformal homeomorphisms and dynamics III: The Teichmüller space of a holomorphic dynamical system. Advances in Mathematics. 1988;**135**: 351-395

[66] Shiga H. Holomorphic families of Riemann surfaces and monodromy. In: Handbook of Teichmller Theory. Vol. IV, 439458, IRMA Lect. Math. Theor. Phys , 19. Zurich: European Mathematical Society; 2014

[67] Sugawa T. Holomorphic motions and quasiconformal extensions. Annales Universitatis Mariae Curie-Sklodowska, Section A. 1999;**53**:239-252

[68] Sugawa T. The universal Teichmller space and related topics. In: Quasiconformal Aappings and Their Applications. New Delhi: Narosa; 2007. pp. 261-289

[69] Bers L, Royden HL. Holomorphic families of injections. Acta Mathematica. 1986;**157**:259-286

[70] Kra I. Automorphic Forms and Kleinian Groups. Reading, MA: Benjamin, Inc.; 1972

[71] Abikoff W. Real analytic theory of Teichmüller spaces. In: Lecture Notes in Mathematics. Vol. 820. Berlin: Springer-Verlag; 1980

[72] Earle CJ, Nag S. Conformally natural reflections in Jordan curves with applications to Teichmüller spaces. In: Drasin D et al, editors. Holomorphic Functions and Moduli II, Mathematical Sciences Reserch Institute Publications. Vol. 11. New York: Springer; 1998. pp. 179-194

[73] Gaier D. Konstruktive Methoden der konformen Abbildung. Berlin: Springer; 1964

[74] Kobayayshi S. Hyperbolic Complex Spaces. New York: Springer; 1998

[75] Lehto O. Univalent Functions and Teichmüller Spaces. New York: Springer-Verlag; 1987

[76] Mitra S. Teichmüller spaces and holomorphic motions. Journal d'Analyse Mathématique. 2000;**81**:1-33

[77] Nehari Z. The Schwarzian derivative and schlicht functions. Bulletin of the American Mathematical Society. 1949; **55**:545-551

[78] Osgood B. Old and new in the Schwarzian derivative. In: Quasiconformal Mappings and Analysis (Ann Arbor, MI, 1975). New York: Springer; 1998. pp. 275-308

[79] Reich E, Strebel K. Extremal quasiconformal mappings with given boundary values. In: Ahlfors LV et al., editors. Contributions to Analysis. New York: Academic Press; 1974. pp. 375-391

[80] Royden HL. A modification of the Neumann-Poincaré method for multiply connected domains. Pacific Journal of Mathematics. 1952;**2**:385-394

[81] Schober G. Semicontinuity of curve functionals. Archive for Rational Mechanics and Analysis. 1969;**53**: 374-376

[82] Strebel K. On the existence of extremal Teichmueller mappings. Journal d'Analyse Mathématique. 1976; **30**:464-480

[83] Sullivan D. Conformal dynamical systems. Geometric dynamics (Rio de Janeiro, 1981). In: Lecture Notes in Mathematics. Vol. **1007**. Berlin: Springer-Verlag; 1983. pp. 725-752

[84] Sullivan D. Quasiconformal homeomorphisms and dynamics. II: Structural stability implies hyperbolicity for Kleinian groups. Acta Mathematica. 1985;**155**:243-260

[85] Sullivan D, Thurston WP. Extending holomorphic motions. Acta Mathematica. 1986;**157**:243-257

[86] Krushkal SL. On Fredholm eigenvalues of unbounded polygons. Siberian Mathematical Journal. 2019; **60**(5):896-901. DOI: 10.1134/ S003744661905001X

[87] Krushkal SL. Quasiconformal reflections across arbitrary planar sets. Scientia, Series A: Mathematical Sciences. 2002;**8**:57-62

[88] Bers L, Kra I, editors. A crash course on Kleinian groups. In: Lecture Notes in Mathematics. Vol. 400. Berlin: Springer; 1974

[89] Krushkal SL. On the question of the structure of the universal Teichmüller space. Soviet Mathematics Doklady. 1989;**38**:435-437

[90] Thurston WP. Zippers and univalent functions. In: Baernstein A II et al., editors. The Bieberbach Conjecture: Proceedings of the Symposium on the Occasion of its Proof. Providence, R.I.: American Mathematical Society; 1986. pp. 185-197

[91] Astala K. Selfsimilar zippers. In: Drasin D et al., editors. Holomorphic Functions and Moduli. Vol. I. New York: Springer; 1988. pp. 61-73

[92] Toki M. On non-starlikeness of Teichmüller spaces. Proceedings of the Japan Academy, Series A. 1993;**69**:58-60

[93] Krushkal SL. On shape of Teichmüller spaces. Journal of Analysis. 2014;**22**:69-76

[94] Ahlfors LV, Weill G. A uniqueness theorem for Beltrami equations. Proceedings of the American Mathematical Society. 1962;**13**:975-978

[95] Krushkal SL. Extremal quasiconformality vs rational approximation. Journal of Mathematical Sciences. 2020;**244**(1):22-35

[96] Krushkal SL. Ukrainian Mathematical Bulletin. 2019;**16**(2):181-199. DOI: 10.1007/s10958-019-04601-6

[97] Krushkal SL. Strengthened Moser's conjecture, geometry of Grunsky coefficients and Fredholm eigenvalues. Central European Journal of Mathematics. 2007;**5**(**3**):551-580

[98] Minda D. The strong form of Ahlfors' lemma. The Rocky Mountain Journal of Mathematics. 1987;**17**:457-461

[99] Krushkal SL. On quasireflections and holomorphic functions. In: Zalcman L, editor. Proceedings of the 1996 Ashkelon Workshop in Complex Function Theory; Israel Mathematical Conference Proceedings. Vol. 11. American Mathematical Society; 1997. pp. 173-185

[100] Kühnau R. Möglichst konforme Spiegelung an einem Jordanbogen auf

der Zahlenebene. Annales Academiae Scientiarum Fennicae. Ser. AI. Mathematica. 1989;**14**:357-367

[101] Kühnau R. Möglichst konforme Jordankurven durch vier Punkte. Revue Roumaine De Mathématiques Pures Et Appliquées. 1991;**36**:383-393

[102] Kühnau R. Zur möglichst konformen Spiegelung. Ann. Univ. Mariae Curie-Sklodowska Lublin, sect. A. 2001;**55**:85-94

[103] Teichmüller O. Ein Verschiebungssatz der quasikonformen Abbildung. Deutsche Mathematik. 1944;**7**:336-343

Chapter 6

Symplectic Geometry and Its Applications on Time Series Analysis

Min Lei

Abstract

This chapter serves to introduce the symplectic geometry theory in time series analysis and its applications in various fields. The basic concepts and basic elements of mathematics relevant to the symplectic geometry are introduced in the second section. It includes the symplectic space, symplectic transformation, Hamiltonian matrix, symplectic principal component analysis (SPCA), symplectic geometry spectrum analysis (SGSA), symplectic geometry mode decomposition (SGMD), and symplectic entropy (SymEn), etc. In addition, it also briefly reviews the applications of symplectic geometry on time series analysis, such as the embedding dimension estimation, nonlinear testing, noise reduction, as well as fault diagnosis. Readers who are familiar with the mathematical preliminaries may omit the second section, i.e. the theory part, and go directly to the third section, i.e. the application part.

Keywords: symplectic geometry, symplectic principal component analysis (SPCA), symplectic geometry spectrum analysis (SGSA), symplectic geometry mode decomposition (SGMD), symplectic entropy (SymEn), chaotic time series, embedding dimension, feature extraction

1. Introduction

From the viewpoint of mathematical systems, the time series observed in physics are usually regarded as coming from the Lagrangian systems, also called the conventional systems. The systems can be analyzed by the conventional Euclidean geometry [1]. However, the systems in practice are usually nonlinear and complex. Thus, a lot of interesting time series in nature are complex due to nonlinear phenomena derived from nonlinear dynamical systems [2]. The nonlinear dynamical systems have been described by Hamiltonian systems and dealt with by using symplectic geometry [3]. Symplectic geometry is an even dimensional geometry living on even dimensional spaces. Different from the conventional Euclidean geometry that measures 1-dimensional lengths and angles, the symplectic geometry studies the metric properties (such as area) and can preserve the system structure in the phase space [4]. Apart from applications on the classical dynamical systems to solve the equation problems, symplectic geometry has been also used on the studies of nonlinear time series [5–8].

According to Takens' embedding theorem, a time series can be reconstructed into an attractor in phase space [9]. The reconstructed attractor is a geometrical object that can reflect the underlying dynamical system. In order to better understand the nature of the underlying system, the attractor and its properties are characterized in the phase space by various mathematical methods, such as dimension, fractal geometry, Lyapunov exponent, entropy and symplectic geometry [1, 5, 10, 11]. For dimension, fractal geometry, Lyapunov exponent, entropy, there are a more extensive discussion with mathematical details in some research literatures [12–15]. Here, we only introduce how to apply symplectic geometry theory to extract the information from the reconstructed attractor and its application on physics, engineering and biomedical engineering.

2. Mathematical fundamental

2.1 Reconstruction of the system dynamics in phase space from a time series

The reconstruction from a time series of observation is the first and most crucial step in nonlinear time series analysis. It is also the basis of applications of symplectic geometry on time series analysis. Takens' embedding theorem allows us to reconstruct an equivalent attractor of the underlying dynamical system by embedding one time series. The theorem proves that the reconstructed attractor has the same dynamical characteristics as the attractor of the original system if the embedding dimension m is sufficiently large. Let a time series of observation $x_1, x_2, \ldots, x_n$. n is the number of samples. The reconstructed attractor can be given in N-dimensional space R^N by the time-delay embedding [5]:

$$\mathbf{X} = (\mathbf{X}_1, \ \mathbf{X}_2, \cdots, \ \mathbf{X}_m) = \begin{pmatrix} x_1 & x_2 & \cdots & x_m \\ x_2 & x_3 & \cdots & x_{m+1} \\ \vdots & \vdots & \ddots & \vdots \\ x_N & x_{N+1} & \cdots & x_n \end{pmatrix}, \qquad (1)$$

where the number of dots in the attractor is $m = n\text{-}N + 1$, the embedding dimension is N. X is also called as the trajectory matrix of the original system in phase space. The corresponding program is given by matlab software as follows:

```
function matrixSignal = signalMatrix(x, N)
% ————Construct data matrix————
%
% Synopsis:
% matrixSignal = signalMatrix(x, N)
%
% Description:
% It constructs a data matrix from a time series as a column vector, i.e., a
% reconstruction attractor.
%
% Input:
% x a time series with the length n.
% N [1x1] Output dimension; N > 1 (default N = dim);
```

```
%
% Ouputs:
% matrixSignal [N x M] a data matrix (M = n-N + 1).
%
if nargin <2, N = 2; end
n = length(x);
M = n-N + 1;

matrixSignal = zeros(N,M);
for i = 1:N
matrixSignal(i,:) = x(i:M + i-1);
end
```

2.2 Hamilton matrix from the reconstructed attractor

In the symplectic spaces, Hamiltonian system is the analysis fundamental for the real physical processes [4, 5]. A real system should be first described by a suitable Hamiltonian system, i.e. an even dimensional matrix. For a time series, its Hamiltonian matrix $\boldsymbol{H}$ can be defined by using its reconstructed attractor $\boldsymbol{X}$.

Definition 2.1 Let $\boldsymbol{X}$ be a d-dimensional matrix in a real number field R^d. The matrix $\overline{X}$can be given by removing the mean values of the columns of the $\boldsymbol{X}$. We define the covariance matrix $\boldsymbol{A}$ of the matrix $\boldsymbol{X}$:

$$\mathbf{A} = \overline{\mathbf{X}} \cdot \overline{\mathbf{X}}^T. \tag{2}$$

Here, $\boldsymbol{A}$ is a $d \times d$ real number matrix.

Definition 2.2 For a $d \times d$ matrix $\boldsymbol{A}$, the Hamiltonian matrix $\boldsymbol{H}$ can be defined:

$$\mathbf{H} = \begin{pmatrix} \mathbf{A} & \mathbf{0} \\ \mathbf{0} & -\mathbf{A}^T \end{pmatrix}. \tag{3}$$

Here, $\boldsymbol{H}$ is a $2d \times 2d$ matrix.

2.3 Mathematical preliminaries in symplectic geometry

Symplectic geometry focuses on the study of area measure in symplectic space $\boldsymbol{R}^{2n}$. Its basic concepts and basic properties are related but different from those of a Euclidean geometry (see **Table 1**).

In Euclidean space, the inner product is denoted as the measure of the length. The unit matrix is $\boldsymbol{I}$, i.e. the main diagonal elements are 1, and the other elements are 0. Corresponding to the unit matrix $\boldsymbol{I}$ in Euclidean space, the unit matrix in symplectic space is defined as the unit symplectic matrix $\boldsymbol{J}$, an even dimensional matrix:

$$\mathbf{J} = \mathbf{J}_{2n} = \begin{bmatrix} 0 & +\mathbf{I}_n \\ -\mathbf{I}_n & 0 \end{bmatrix}, \tag{4}$$

The properties of the matrix $\boldsymbol{J}$ have:

$$|\mathbf{J}| = 1, \tag{5}$$

$$\mathbf{J}^2 = -\mathbf{I}, \tag{6}$$

$$\mathbf{J}^T = \mathbf{J}^{-1} = -\mathbf{J}, \tag{7}$$

Geometry space	Symplectic space	Euclidean space
Space dimension	$2n$-dimension	n-dimension
Unit matrix	unit symplect matrix: $J_{2n} = \begin{pmatrix} \mathbf{0} & +I_n \\ -I_n & \mathbf{0} \end{pmatrix}$.	unit matrix: $I_n = \underbrace{\begin{pmatrix} 1 & 0 & \cdots & 0 \\ 0 & 1 & \ddots & \vdots \\ \vdots & \ddots & \ddots & 0 \\ 0 & \cdots & 0 & 1 \end{pmatrix}}_{n}$.
Determinant of unit matrix	$\|J_{2n}\| = 1$	$\|I_n\| = 1$
Product calculation	symplectic inner product $<x, y>$ $\langle x,\ y\rangle = (x,\ J_{2n}y) = x^T J_{2n} y$	Inner product (x, y) $(x,\ y) = (x,\ I_n y) = x^T I_n y = x^T y$
Calculation measure	area	length
Orthogonality	$\langle x,\ y\rangle = x^T J_{2n} y = 0$.	$(x,\ y) = x^T I_n y = x^T y = 0$
Space basis	Adjoint symplectic orthonormal basis $Q = \{x_1, x_2, \ldots, x_m, y_1, y_2, \ldots, y_m\}$, $m \le n$; when determinant $\|Q\| = 1$, the basis Q is normal.	Orthogonal basis $W = \{x_1, x_2, \ldots, x_m\}$, $m \le n$; when $\|W\| = 1$, the basis W is normal.
Orthogonal matrix	Symplectic matrix S $S^T J S = J$.	Orthogonal matrix W $W^T I W = W^T W = I$.
Analysis matrix	Hamiltonian matrix H $H^T = JHJ$.	Symmetry matrix A $A^T = A = IAI$.
Matrix transformation	Hamiltonian transformation $\langle x, Hy\rangle = \langle y, Hx\rangle$.	Symmetry transformation $(x, Ay) = (y, Ax)$.
Eigenvalues of the matrix	The eigenvalues of H are $\pm\mu$.	The eigenvalues μ of A are real.
Eigenvectors of the matrix	The eigenvectors of H are symplectic orthogonal.	The eigenvectors of A are orthogonal.

Table 1.
The comparison between symplectic geometry and Euclidean geometry.

$$\mathbf{J}\mathbf{J}^{-1} = \mathbf{J}^{-1}\mathbf{J} = \mathbf{I}. \tag{8}$$

Definition 2.3 For any two n-dimensional vectors $x_{2n \times 1}$ and $y_{2n \times 1}$, the normal symplectic inner product is defined by using the inner product of Euclidean space:

$$\langle \mathbf{x},\ \mathbf{y}\rangle = (\mathbf{x},\ \mathbf{J}_{2n}\mathbf{y}) = \sum_{i=1}^{n} \left(x_i y_{n+i} - x_{n+i} y_i\right) = \mathbf{x}^T \mathbf{J}_{2n} \mathbf{y}. \tag{9}$$

The normal symplectic inner product is also denoted briefly as the symplectic inner product in a real vector space R^{2n}. When $n = 1$, there is:

$$\mathbf{J}_2 = \begin{bmatrix} 0 & 1 \\ -1 & 0 \end{bmatrix}, \tag{10}$$

$$\langle \mathbf{x},\ \mathbf{y} \rangle = (\mathbf{x},\ \mathbf{J}_2\mathbf{y}) = \mathbf{x}^T\mathbf{J}_2\mathbf{y} = \begin{vmatrix} x_1 & y_1 \\ x_2 & y_2 \end{vmatrix} = x_1y_2 - x_2y_1. \tag{11}$$

The symplectic inner product is a bilinear antisymmetric nonsingular cross product. In symplectic space, the length of any vectors is equal to 0. But there exists the concept of symplectic orthogonal cross-course.

Definition 2.4 Let $\boldsymbol{x}$ and $\boldsymbol{y}$ be a $2n$-dimensional real vector. If their symplectic inner product is equal to zero, i.e.:

$$\langle \mathbf{x},\ \mathbf{y} \rangle = \mathbf{x}^T\mathbf{J}\mathbf{y} = 0, \tag{12}$$

then $\boldsymbol{x}$ and $\boldsymbol{y}$ are symplectic orthogonal. Otherwise, they are called as symplectic adjoint.

Definition 2.5 If a vector set $\{\boldsymbol{x}_1, \boldsymbol{x}_2, \dots, \boldsymbol{x}_m, \boldsymbol{y}_1, \boldsymbol{y}_2, \dots, \boldsymbol{y}_m\}$ in the real symplectic space $\boldsymbol{R}^{2n}$ $(m \le n)$is an adjoint symplectic orthonormal vector set, then the vectors $\boldsymbol{x}_i$ and $\boldsymbol{y}_i$ $(i = 1, \dots, m,\ \boldsymbol{x}_i \in \boldsymbol{R}^{2n},\ \boldsymbol{y}_i \in \boldsymbol{R}^{2n})$ satisfy

$$\left\langle \mathbf{x}_i,\ \mathbf{y}_j \right\rangle = \mathbf{x}_i^T\mathbf{J}_{2n}\mathbf{y}_j = \begin{cases} a_{ii} \neq 0, & i = j \\ 0, & i \neq j \end{cases}, \tag{13}$$

$$\left\langle \mathbf{x}_i,\ \mathbf{x}_j \right\rangle = 0, \tag{14}$$

$$\left\langle \mathbf{y}_i,\ \mathbf{y}_j \right\rangle = 0, \tag{15}$$

where $i, j = 1, 2, \dots, m$. It is called as an adjoint symplectic orthonormal basis in the $2n$-dimensional symplectic space. If $a_{ii} = 1$, the vector set $\{\boldsymbol{x}_1, \boldsymbol{x}_2, \dots, \boldsymbol{x}_m, \boldsymbol{y}_1, \boldsymbol{y}_2, \dots, \boldsymbol{y}_m\}$ is a normal adjoint symplectic orthonormal vector set (a normal adjoint symplectic orthonormal basis in the space $\boldsymbol{R}^{2n}$).

The orthogonal of the Euclidean space is different from the symplectic orthogonal. If vectors $\boldsymbol{x}$ and $\boldsymbol{y}$ in the space $\boldsymbol{R}^n$ are orthonormal, then they satisfy:

$$(\mathbf{x},\ \mathbf{y}) = \mathbf{x}^T\mathbf{y} = 0, \tag{16}$$

where $\boldsymbol{x} \neq \boldsymbol{y}$.

If a vector set $\{\boldsymbol{x}_1, \boldsymbol{x}_2, \dots, \boldsymbol{x}_m\} \in \boldsymbol{R}^n$ is an orthonormal vector set, then any two vectors in the set satisfy:

$$(\mathbf{x}_i,\ \mathbf{x}_j) = 0, \tag{17}$$

where $i, j = 1, 2, \dots, m,\ i \neq j$. Eq. (17) is similar to Eqs. (14) and (15). In the n-dimensional Euclidean space, the set $\{\boldsymbol{x}_i\}$ is denoted as an orthonormal basis. If $||\boldsymbol{x}_i|| = 1$, the orthonormal basis is a normal orthonormal basis.

Theorem 2.1 Let $\{\alpha_i\}$ be a normal adjoint symplectic orthonormal basis in a $2n$-dimensional symplectic space Φ. Let the coordinates of any vectors $\boldsymbol{\beta}$ and $\boldsymbol{\gamma}$ in Φ be $\{x_1, x_2, \dots x_n, x_{n+1}, \dots, x_{2n}\}^{\mathrm{T}}$ and $\{y_1, y_2, \dots y_n, y_{n+1}, \dots, y_{2n}\}^{\mathrm{T}}$, respectively. Referring to the basis $\{\alpha_i\}$, the coordinates can be described as:

$$x_i = \langle \boldsymbol{\beta},\ \alpha_{n+i} \rangle,\ x_{n+i} = -\langle \boldsymbol{\beta},\ \alpha_i \rangle, y_i = \langle \boldsymbol{\gamma},\ \alpha_{n+i} \rangle,\ y_{n+i} = -\langle \boldsymbol{\gamma},\ \alpha_i \rangle, \tag{18}$$

where $i = 1, 2, \dots, n$. Then the symplectic inner product of $\boldsymbol{\beta}$ and $\boldsymbol{\gamma}$ is as follows:

$$\langle \boldsymbol{\beta},\ \boldsymbol{\gamma} \rangle = \sum_{i=1}^{n} \left(x_i y_{n+i} - x_{n+i} y_i \right) = \mathbf{x}^T\mathbf{J}_{2n}\mathbf{y}. \tag{19}$$

Thus, the symplectic inner product operation is transformed to the matrix operation of ordinary vectors or matrices by applying a normal adjoint symplectic orthonormal basis.

Definition 2.6 Let S is a $2n \times 2n$ matrix, if S satisfies:

$$\mathbf{JSJ}^{-1} = \mathbf{S}^{-T}, \text{or } \mathbf{S}^T\mathbf{JS} = \mathbf{J}, \tag{20}$$

then S is a symplectic matrix and the determinant $|S| = 1$ or -1. Meanwhile, the inverse matrix and the transpose matrix of a symplectic matrix are a symplectic matrix, respectively. The symplectic matrix S is similar to an orthogonal matrix W in Euclidean space, like Eq. (20):

$$\mathbf{W}^T\mathbf{IW} = \mathbf{W}^T\mathbf{W} = \mathbf{I}. \tag{21}$$

Theorem 2.2 The product of sympletcic matrixes is also a symplectic matrix.

Proof:

Let S_i (i = 1, 2, ..., n) be a symplectic matrix. The product matrix M:

$$\mathbf{M} = \prod_{i=1}^{n} \mathbf{S}_i. \tag{22}$$

According to the above definition of symplectic matrix, there are:

$$\mathbf{JS}_i\mathbf{J}^{-1} = \mathbf{S}_i^{-T}, \quad i = 1, 2, \cdots, n \tag{23}$$

$$\mathbf{J}^{-1}\mathbf{J} = \mathbf{I}, \tag{24}$$

$$\begin{aligned} \mathbf{JMJ}^{-1} &= \mathbf{J}\left(\prod_{i=1}^{n} \mathbf{S}_i\right)\mathbf{J}^{-1} \\ &= \mathbf{J}(\mathbf{S}_1\mathbf{S}_2\cdots\mathbf{S}_n)\mathbf{J}^{-1} \\ &= \mathbf{JS}_1\mathbf{J}^{-1}\mathbf{JS}_2\mathbf{J}^{-1}\mathbf{J}\cdots\mathbf{J}^{-1}\mathbf{JS}_n\mathbf{J}^{-1} \\ &= (\mathbf{JS}_1\mathbf{J}^{-1})(\mathbf{JS}_2\mathbf{J}^{-1})(\mathbf{J}\cdots\mathbf{J}^{-1})(\mathbf{JS}_n\mathbf{J}^{-1}), \\ &= \mathbf{S}_1^{-T}\mathbf{S}_2^{-T}\cdots\mathbf{S}_n^{-T} \\ &= (\mathbf{S}_1\mathbf{S}_2\cdots\mathbf{S}_n)^{-T} \\ &= \mathbf{M}^{-T} \end{aligned} \tag{25}$$

Thus, the product of symplectic matrixes is also a symplectic matrix.

Definition 2.7 If a $2n \times 2n$ matrix H is a Hamiltonian matrix, then the matrix H satisfies the following properties:

$$\mathbf{JHJ}^{-1} = -\mathbf{H}^T, \mathbf{JHJ} = \mathbf{H}^T, \text{or } (\mathbf{JH})^T = \mathbf{JH}, \tag{26}$$

$$\langle \mathbf{x}, \mathbf{Hy} \rangle = \langle \mathbf{y}, \mathbf{Hx} \rangle, \tag{27}$$

where x and y are $2n$-dimensional vectors. In other words, if an even-dimensional matrix H satisfies these properties above, the matrix H is a Hamiltonian matrix. In Euclidean space, a symmetric matrix A is similar to a Hamilitonian matrix H, like Eqs. (26) and (27):

$$\mathbf{IAI} = \mathbf{A} = \mathbf{A}^T, \tag{28}$$

$$(\mathbf{x}, \mathbf{Ay}) = (\mathbf{y}, \mathbf{Ax}). \tag{29}$$

Theorem 2.3 Let a matrix A be a $n \times n$ real number matrix, if it can be built into a $2n \times 2n$ matrix H in symplectic space in the following pattern:

$$\begin{pmatrix} \mathbf{A} & \mathbf{0} \\ \mathbf{0} & -\mathbf{A}^T \end{pmatrix}. \tag{30}$$

Then the matrix H is a Hamilton matrix.

Proof:

Let $\mathbf{H} = \begin{pmatrix} \mathbf{A} & \mathbf{0} \\ \mathbf{0} & -\mathbf{A}^T \end{pmatrix}$, then

$$\begin{aligned} \mathbf{JHJ}^{-1} &= \mathbf{J}\begin{pmatrix} \mathbf{A} & \mathbf{0} \\ \mathbf{0} & -\mathbf{A}^T \end{pmatrix}\mathbf{J}^{-1} \\ &= \begin{pmatrix} 0 & \mathbf{I}_n \\ -\mathbf{I}_n & 0 \end{pmatrix}\begin{pmatrix} \mathbf{A} & \mathbf{0} \\ \mathbf{0} & -\mathbf{A}^T \end{pmatrix}\begin{pmatrix} 0 & \mathbf{I}_n \\ -\mathbf{I}_n & 0 \end{pmatrix}^{-1} \\ &= \begin{pmatrix} -\mathbf{A}^T & \mathbf{0} \\ \mathbf{0} & \mathbf{A} \end{pmatrix} \\ &= -\begin{pmatrix} \mathbf{A} & \mathbf{0} \\ \mathbf{0} & -\mathbf{A}^T \end{pmatrix}^T \\ &= -\mathbf{H}^T \end{aligned}, \tag{31}$$

where J is the $2n \times 2n$ unit symplectic matrix. In terms of Definition 2.7, the matrix H is a $2n \times 2n$ Hamiltonian matrix.

Theorem 2.4 Let a $2n \times 2n$ matrix H be a Hamiltonian matrix. Then its properties keep unchanged at symplectic similar transform. That is, a Hamiltonian matrix H through a series of symplectic similar transforms is still a Hamiltonian matrix.

Proof:

According to Definition 2.6, let the matrix S be a symplectic transform matrix. Then, the inverse matrix S^{-1} is also a symplectic matrix. For a Hamiltonian matrix H, let SHS^{-1} be the matrix M under the symplectic similar transformation of the matrices S and S^{-1}. Thus,

$$\begin{aligned} \mathbf{J}(\mathbf{M})\mathbf{J}^{-1} &= \mathbf{J}(\mathbf{SHS}^{-1})\mathbf{J}^{-1} \\ &= (\mathbf{JSJ}^{-1})(\mathbf{JHJ}^{-1})(\mathbf{JS}^{-1}\mathbf{J}^{-1}) \\ &= \mathbf{S}^{-T}(-\mathbf{H}^T)\mathbf{S}^T \\ &= -(\mathbf{SHS}^{-1})^T \\ &= -\mathbf{M}^T \end{aligned}. \tag{32}$$

Therefore, M is also a Hamiltonian matrix. Moreover, the matrix M is similar to the matrix H. Therefore, the Hamiltonian matrix H can keep unchanged at symplectic similar transform in symplectic space.

The eigenvalues of a Hamiltonian matrix have the specific characteristics of the Hamiltonian matrix. However, the eigenvalues may be complex or repeated eigenvalues. In order to obtain the real eigenvalues of a Hamiltonian matrix $\boldsymbol{H}$, symplectic *QR* decomposition method is applied to deal with the Hamiltonian $\boldsymbol{H}$:

1. Let a $2n \times 2n$ matrix $\boldsymbol{H}$ be $(\boldsymbol{A}^{\mathrm{T}}\ \boldsymbol{G};\ \boldsymbol{F}\ -\boldsymbol{A})$, then

$$\mathbf{N} = \mathbf{H}^2 = \begin{pmatrix} \mathbf{A}^T & \mathbf{G} \\ \mathbf{F} & -\mathbf{A} \end{pmatrix}^2, \tag{33}$$

2. Build a $2n \times 2n$ symplectic matrix $\mathbf{Q}$ and satisfy:

$$\mathbf{Q}^T\mathbf{N}\mathbf{Q} = \begin{pmatrix} \mathbf{B} & \mathbf{R} \\ \mathbf{0} & \mathbf{B}^T \end{pmatrix}, \tag{34}$$

$$\mathbf{B} = \begin{pmatrix} b_{11} & b_{12} & \cdots & \cdots & b_{1n} \\ b_{21} & b_{22} & \cdots & \cdots & b_{2n} \\ 0 & b_{32} & \ddots & \ddots & \vdots \\ \vdots & \ddots & \ddots & \ddots & \vdots \\ 0 & \cdots & 0 & b_{nn-1} & b_{nn} \end{pmatrix}. \tag{35}$$

Here $\boldsymbol{B}$ is an upper Hessenberg matrix. Besides, the matrix $\mathbf{Q}$ can be a $2n \times 2n$ Householder matrix.

3. Use the symplectic *QR* decomposition method to obtain eigenvalues:

$$\mu(B) = \{\mu_1,\ \mu_2,\ \cdots,\ \mu_n\}. \tag{36}$$

4. The eigenvalues of the Hamiltonian matrix $\boldsymbol{H}$ with multiplicity n are $\lambda_i = \sqrt{\mu_i}$, $i = 1, 2, \ldots, n$; $\lambda_{n+i} = -\lambda_i$ is also an eigenvalue with multiplicity n.

In symplectic space, the symplectic *QR* decomposition method allows the primary $2n$-dimensional space transform into n dimensional space to resolve the eigenvalues of the Hamiltonian $\boldsymbol{H}$, where the matrix $\mathbf{Q}$ is a symplectic unitary matrix. Thus, the consuming time of the calculation is only one fourth the number of floating-point operations. In general, one makes use of a Householder matrix instead of the matrix $\mathbf{Q}$.

Theorem 2.5 If a $2n \times 2n$ matrix $\mathbf{Q}$ is a Householder matrix, then the matrix $\mathbf{Q}$ is a symplectic unitary matrix.

Proof:

Let a Householder matrix $\mathbf{Q}$

$$\mathbf{Q} = \mathbf{Q}(k, \omega) = \begin{pmatrix} \mathbf{P} & 0 \\ 0 & \mathbf{P} \end{pmatrix}, \tag{37}$$

$$\mathbf{P} = \mathbf{I}_n - \frac{2\boldsymbol{\omega}\boldsymbol{\omega}^*}{\boldsymbol{\omega}^*\boldsymbol{\omega}}, \tag{38}$$

$$\boldsymbol{\omega} = (0, \cdots, 0, \omega_k, \cdots, \omega_n)^T \neq 0, \tag{39}$$

where, '*' means the conjugate transposition. Then, there is

$$\mathbf{P}^* = \mathbf{P}, \tag{40}$$

$$\begin{aligned}\mathbf{P}^*\mathbf{P} &= \mathbf{P}^2 \\ &= \left(\mathbf{I}_n - \frac{2\omega\omega^*}{\omega^*\omega}\right)\left(\mathbf{I}_n - \frac{2\omega\omega^*}{\omega^*\omega}\right), \\ &= \mathbf{I}_n\end{aligned} \tag{41}$$

$$\begin{aligned}\mathbf{Q}^*\mathbf{J}\mathbf{Q} &= \begin{pmatrix}\mathbf{P} & \mathbf{0}\\ \mathbf{0} & \mathbf{P}\end{pmatrix}^*\begin{pmatrix}\mathbf{0} & \mathbf{I}_n\\ -\mathbf{I}_n & \mathbf{0}\end{pmatrix}\begin{pmatrix}\mathbf{P} & \mathbf{0}\\ \mathbf{0} & \mathbf{P}\end{pmatrix} \\ &= \begin{pmatrix}\mathbf{0} & \mathbf{P}^*\mathbf{P}\\ -\mathbf{P}^*\mathbf{P} & \mathbf{0}\end{pmatrix}, \\ &= \begin{pmatrix}\mathbf{0} & \mathbf{I}_n\\ -\mathbf{I}_n & \mathbf{0}\end{pmatrix} \\ &= \mathbf{J}\end{aligned} \tag{42}$$

$$\begin{aligned}\mathbf{Q}^*\mathbf{Q} &= \begin{pmatrix}\mathbf{P} & \mathbf{0}\\ \mathbf{0} & \mathbf{P}\end{pmatrix}^*\begin{pmatrix}\mathbf{P} & \mathbf{0}\\ \mathbf{0} & \mathbf{P}\end{pmatrix} \\ &= \begin{pmatrix}\mathbf{P}^*\mathbf{P} & \mathbf{0}\\ \mathbf{0} & \mathbf{P}^*\mathbf{P}\end{pmatrix}, \\ &= \begin{pmatrix}\mathbf{I}_n & \mathbf{0}\\ \mathbf{0} & \mathbf{I}_n\end{pmatrix} \\ &= \mathbf{I}_{2n}\end{aligned} \tag{43}$$

Therefore, the Householder matrix $\boldsymbol{Q}$ is a symplectic unitary matrix.

2.4 Mathematical fundamental on applications

2.4.1 Symplectic geometry spectrums of the reconstructed attractor from a time series

In symplectic space, the reconstructed attractor can keep its properties unchanged [5, 6]. Its symplectic geometry spectrums can be given by the symplectic geometry theory above. On the basis of Section 2.1 and 2.2, one can build a Hamiltonian matrix $\boldsymbol{M}$ from a time series of the observation. Due to the structure characteristics of the matrix $\boldsymbol{M}$, its eigenvalues can be calculated by the $2n$-dimensional symplectic space reducing into n-dimensional space. In terms of Theorem 1.5, a $2n \times 2n$ symplectic Householder matrix $\boldsymbol{Q}$ can be constructed. The matrix $\boldsymbol{P}$ in the matrix $\boldsymbol{Q}$ can be calculated by the matrix $\boldsymbol{A}$ in the matrix $\boldsymbol{M}$. The specific steps are as follows:

1. Let $\boldsymbol{A}$ be

$$\begin{aligned}\boldsymbol{A} &= \left(\begin{array}{c|ccc} a_{11} & a_{12} & \cdots & a_{1n} \\ \hline a_{21} & a_{22} & \cdots & a_{2n} \\ \vdots & \vdots & \ddots & \vdots \\ a_{n1} & a_{n2} & \cdots & a_{nn}\end{array}\right). \\ &= \left(\begin{array}{c|c} a_{11} & \boldsymbol{A}_{12}^{(1)} \\ \hline \boldsymbol{\alpha}_{21}^{(1)} & \boldsymbol{A}_{22}^{(1)}\end{array}\right)\end{aligned} \tag{44}$$

If the vector $\boldsymbol{\alpha}_{21}^{(1)} \neq 0$, set S(1) be the first column vector of A:

$$\mathbf{S}^{(1)} = \begin{pmatrix} a_{11}^{(1)} \\ a_{21}^{(1)} \\ \vdots \\ a_{n1}^{(1)} \end{pmatrix} = \begin{pmatrix} a_{11} \\ a_{21} \\ \vdots \\ a_{n1} \end{pmatrix}, \tag{45}$$

then, there is:

$$\alpha^{(1)} = \left\|\mathbf{S}^{(1)}\right\|_2, \tag{46}$$

$$\rho^{(1)} = \left\|\mathbf{S}^{(1)} - \alpha^{(1)}\mathbf{E}^{(1)}\right\|_2, \tag{47}$$

$$\boldsymbol{\omega}^{(1)} = \frac{\mathbf{S}^{(1)} - \alpha^{(1)}\mathbf{E}^{(1)}}{\rho^{(1)}}, \tag{48}$$

where $\boldsymbol{E}^{(1)}$ = (1, 0, ..., 0)^T is a $n \times 1$ unit column vector.
Then, the elementary reflective matrix $\boldsymbol{P}^{(1)}$ can be calculated:

$$\mathbf{P}^{(1)} = \mathbf{I} - 2\boldsymbol{\omega}^{(1)}\left(\boldsymbol{\omega}^{(1)}\right)^T. \tag{49}$$

So, there is

$$\begin{aligned} \mathbf{A}^{(2)} &= \mathbf{P}^{(1)}\mathbf{A} \\ &= \begin{pmatrix} \sigma_1 & a_{12}^{(2)} & \cdots & a_{1n}^{(2)} \\ 0 & a_{22}^{(2)} & \cdots & a_{2n}^{(2)} \\ \vdots & \vdots & \ddots & \vdots \\ 0 & a_{n2}^{(2)} & \cdots & a_{nn}^{(2)} \end{pmatrix}. \end{aligned} \tag{50}$$

Continue to deal with $\boldsymbol{A}^{(2)}$ by repeating the above steps, let $\boldsymbol{S}^{(2)}$ be

$$\mathbf{S}^{(2)} = \begin{pmatrix} 0 \\ a_{22}^{(2)} \\ \vdots \\ a_{n2}^{(2)} \end{pmatrix}. \tag{51}$$

Then,

$$\alpha^{(2)} = \left\|\mathbf{S}^{(2)}\right\|_2, \tag{52}$$

$$\rho^{(2)} = \left\|\mathbf{S}^{(2)} - \alpha^{(2)}\mathbf{E}^{(2)}\right\|_2, \tag{53}$$

$$\boldsymbol{\omega}^{(2)} = \frac{\mathbf{S}^{(2)} - \alpha^{(2)}\mathbf{E}^{(2)}}{\rho^{(2)}}, \tag{54}$$

where $\boldsymbol{E}^{(2)}$ = (0, 1, 0, ..., 0)^T is a $n \times 1$ unit column vector.
Then, the elementary reflective matrix $\boldsymbol{P}^{(2)}$ can be calculated:

$$\mathbf{P}^{(2)} = \mathbf{I} - 2\boldsymbol{\omega}^{(2)}\left(\boldsymbol{\omega}^{(2)}\right)^T. \tag{55}$$

Thus, we can get $\boldsymbol{A}^{(3)}$ with all zeros elements except the first and second non-zero elements:

$$\mathbf{A}^{(3)} = \mathbf{P}^{(2)}\mathbf{A}^{(2)} = \begin{pmatrix} \sigma_1 & a_{12}^{(3)} & a_{13}^{(3)} & \cdots & a_{1n}^{(3)} \\ 0 & \sigma_2 & a_{23}^{(3)} & \cdots & a_{2n}^{(3)} \\ 0 & 0 & a_{33}^{(3)} & \vdots & \vdots \\ \vdots & \vdots & \vdots & \ddots & \vdots \\ 0 & 0 & a_{n3}^{(3)} & \cdots & a_{nn}^{(3)} \end{pmatrix}. \tag{56}$$

Repeat the same steps until $\boldsymbol{A}^{(n)}$ becomes an upper triangle matrix, one can construct a Householder matrix $\boldsymbol{P}$ as follows:

$$\mathbf{P} = \mathbf{P}^{(n)}\mathbf{P}^{(n-1)}\dots\mathbf{P}^{(1)}. \tag{57}$$

Thus, a symplectic Householder matrix $\boldsymbol{Q}$ can be built to make the Hamiltonian matrix $\boldsymbol{M}$ transform as an upper Hessenberg matrix, namely:

$$\mathbf{QMQ}^T = \begin{pmatrix} \mathbf{P} & \mathbf{0} \\ \mathbf{0} & \mathbf{P} \end{pmatrix}\begin{pmatrix} \mathbf{A} & \mathbf{0} \\ \mathbf{0} & -\mathbf{A}^T \end{pmatrix}\begin{pmatrix} \mathbf{P} & \mathbf{0} \\ \mathbf{0} & \mathbf{P} \end{pmatrix}^T = \begin{pmatrix} \mathbf{PAP}^T & \mathbf{0} \\ \mathbf{0} & -\mathbf{PA}^T\mathbf{P}^T \end{pmatrix} = \begin{pmatrix} \mathbf{B} & \mathbf{0} \\ \mathbf{0} & -\mathbf{B}^T \end{pmatrix}. \tag{58}$$

$$\mu(\mathbf{A}) = \mu(\mathbf{B}), \tag{59}$$

where μ means the eigenvalue. The matlab program is as follow:

```
function [P, R] = householder (A)
% ————Solve Householder Transform Matrix————
%
% Synopsis:
% [P, R] = householder (A)
%
% Description:
% It solves a Householder matrix from a data matrix, i.e., a
% reconstruction attractor.
%
% Input:
% A [mRow x mCol] a data matrix.
%
```

```
% Ouputs:
% P [mRow x mRow] a Householder matrix
% R [mRow x mCol] an upper triangle matrix

[mRow, mCol] = size(A);
if mRow>mCol
    A = A';
    [mRow, mCol] = size(A);
end
I_matrix = eye(mRow);
m = min([mRow, mCol]);
p = I_matrix;
for i = 1:m.
    S = A(:,i);
if i > 1.
S(1:i-1) = 0;
end
alpha = sqrt(S'*S);
    delta1 = S-alpha*I_matrix(:,i);
delta = sqrt(delta1'*delta1);
if delta==0
delta = eps;
end
omega = delta1/delta;
    p = I_matrix-2*omega*omega';
    A = p*A;
    P = p*P;
end
R = A;
return
```

For the attractor matrix $\boldsymbol{X}$ of a time series, its symplectic geometry spectrums $\boldsymbol{SGS}$ are calculated by the eigenvalues of the $\boldsymbol{A}$ in descending order, that is:

$$SGS_i = \log\left(\frac{\sigma_i}{tr(\sigma_i)}\right), \tag{60}$$

$$\sigma = \mu^2(\mathbf{X}) = \mu(\mathbf{A}), \sigma_1 = \mu_{\max}^2, \quad \cdots, \quad \sigma_n = \mu_{\min}^2, \tag{61}$$

where $i = 1, \ldots, n$. n is the dimension of the attractor $\boldsymbol{X}$.

2.4.2 *Embedding dimension estimate of the reconstructed attractor from a time series*

To estimate the embedding dimension is usually the first step of nonlinear analysis [5]. For a time series, it is important to resolve a suitable embedding dimension of the observed system. Due to the measure-preserving charactistic of symplectic geometry, symplectic geometry spectrums can be used to estimate the embedding dimension of the system from a time series. With the increase of the dimension n in Eq. (61), the change of the symplectic geometry spectrums SGS in Eq. (60) tends to be flat at $i = d$ $(i \in (1,n))$and enters the noise floor area, $SGS_1 > SGS_2 > \ldots > SGS_d > > SGS_{d+1} \geq \ldots \geq SGS_n$,. That is, the eigenvalues exist $\sigma_1 > \sigma_2 > \ldots > \sigma_d > > \sigma_{d+1} \geq \ldots \geq \sigma_n$, then d is defined as the embedding dimension of the time series for the reconstruction system.

2.4.3 Symplectic entropy (SymEn) of a time series

Symplectic entropy(SymEn) is a kind of entropy measure for a dynamic system in symplectic space [16]. Based on the symplectic geometry spectrums, the SymEn measures the energy distribution in symplectic space of a dynamic system from a time series. The distribution of the energy of the system is described by the eigenvalues $\boldsymbol{\sigma}$ in the relevant symplectic orthonormal bases of the symplectic space. In each base direction, the probability of the energy distribution can be given as follows:

$$p_i = \frac{\sigma_i}{\sum_{i=1}^{n} \sigma_i}, \tag{62}$$

where i denotes the ith base direction in the symplectic space, $\sum_{i=1}^{n} p_i = 1$, $0 \le p_i \le 1$.

Then,

$$SymEn = -\sum_{i=1}^{n} p_i \log(p_i). \tag{63}$$

The matlab program is as follows:

```
function SymEn = SymplecticEntropy(A)
   [Q, R] = householder(A);
delta = diag(R);
sum_delta = sum(delta);
   p = delta./sum_delta;
SymEn = −sum(p.*log(p));
Return
```

The SymEn value represents the uncertainty of the entropy about the underlying probability distribution of a dynamic system in symplectic space, called Symplectic Entropy.

2.4.4 Symplectic principal component analysis (SPCA) of a time series

Symplectic principal component analysis (SPCA) is a kind of principal component analysis (PCA) to map the dynamic system from a time series into the symplectic space [17]. Due to the preserving-measure nature of symplectic geometry, symplectic principal components elucidate the dominant features of a time series for an underlying system. The principal components corresponding to larger eigenvalues capture the key relationship between the variables in symplectic space. The components corresponding to smaller eigenvalues are regarded to relate primarily to the less important components or noise in the time series. The analysis of eigenvalues are also called as the symplectic geometry spectrums analysis (SGSA) [6, 18, 19]. The corresponding components are also regarded as symplectic geometry mode decomposition (SGMD) [7, 8, 20, 21]. According to the symplectic geometry spectrums above, if the number of the chosen symplectic principal components is k, the corresponding principal eigenvector matrix $\boldsymbol{p}$ can be constructed by using the first k eigenvectors of the matrix $\boldsymbol{P}$ in the matrix $\boldsymbol{Q}$. The corresponding principal eigenvalues are the first k eigenvalues in the symplectic geometry spectrum. If $k = n$, $p = P$. Otherwise, $p \subset P$. Then the reestimated attractor matrix $\hat{\mathbf{X}} = \mathbf{p}(\mathbf{p}^T\mathbf{X})$, where $p^T X$ is defined the transformation coefficient matrix S. If p_i is the ith eigenvector in P corresponding to the ith eigenvalue σ_i in the symplectic geometry spectrum, S_i will be the ith principal component coefficients, or called the projection of the p_ith direction in the symplectic space:

$$\mathbf{S}_i = \mathbf{p}_i^T\mathbf{X} = \mathbf{X}^T\mathbf{p}_i. \tag{64}$$

The corresponding p_ith principal component matrix $\hat{\mathbf{X}}_i$ is given as follows:

$$\hat{\mathbf{X}}_i = \mathbf{p}_i\mathbf{S}_i. \tag{65}$$

Then, the reestimated attractor matrix is equal to the sum of $\hat{\mathbf{X}}_i$, $i = 1, \ldots, n$.

$$\hat{\mathbf{X}} = \sum_{i=1}^{n} \hat{\mathbf{X}}_i. \tag{66}$$

The reestimated time series x^r is equal to the sum of each principal component, i.e. the sum of projections in different directions. If $i = 1$, the reestimated time series is a reduced noise data based on the first principal component.

3. Applications

Symplectic geometry theory has been applied to deal with a time series in fields of physics, engineering, biomedical engineering [6–8, 11, 16–24], since Lei et al. (2002) first proposed a symplectic geometry method to estimate the appropriate embedding dimension from a time series [5]. Here, we will introduce four research cases in terms of the above theorem and properties of symplectic geometry for the time series analysis.

Case 1: Embedding dimension estimation for Lorenz chaotic time series [5].

Lorenz chaotic system was accidentally discovered by Edward Norton Lorenz [25], an American meteorologist, in 1963 when he was studying weather forecast, and was known as the first chaotic attractor. Since then, people began to study chaos, a random-like phenomenon. Lorenz chaotic time series x comes from Lorenz chaotic system, which is a three-dimensional dynamical system as follows [5]:

$$\begin{aligned} \dot{x} &= \sigma(y - x) \\ \dot{y} &= \gamma x - y - xz \\ \dot{z} &= -bz + xy \end{aligned}, \tag{67}$$

where $\sigma = 10$, $b = 8/3$, $\gamma = 28$. The state variable x is chosen as the analyzed data. The sampling interval is 0.005. The length n is 1000 points.

The attractor reconstructed from Lorenz chaotic time series x can reflect the Lorenz system. Here, the dimension of the reconstructed attractor is estimated by the above symplectic geometry method. Let the embedding dimension d be 3: 5: 23, where i = 1: d. The matlab program is as follows:

```
% Compute a Lorenz chaotic time series
% Example:
```

```
% state = [5 5 5];
% Ts = 0.005;
% N = 10000;
% y = calculate_lorenz(state, Ts, N);
% x = y(:,1);

function y = calculate_lorenz(state, Ts, N).

if nargin <1
state = [5 5 5];
Ts = 0.005;
   N = 10000;
end
if nargin == 1
Ts = 0.005;
   N = 10000;
end
if nargin == 2
   N = 10000;
end
% set time span with specific times for the solution
T = 0:Ts:N*Ts;
% set a scalar relative error tolerance 'RelTol' (1e-3 by default).
% and a vector of absolute error tolerances 'AbsTol' (all components 1e-6% by
default).
options = odeset('RelTol',1e-4,'AbsTol',[1e-4 1e-4 1e-5]);
% solve Lorenz chaotic system
[t,y] = ode45('lorenzeq1',T,state,options);
return
function ydot = lorenzeq(t,y)
% Lorenz equation
b = 8/3;
r = 28;
delta = 10;
A = [−delta delta 0;r − 1 -y(1);y(2) 0 -b];
ydot = A*y;
return
```

```
% Calculate the embedding dimension.
state = [5 5 5];
Ts = 0.005;
N = 10000;
y = calculate_lorenz(state, Ts, N);
x = y(:,1);
figure.
for N = 3:5:23
   X = signalMatrix(x,N);
   A = X*X';
   [Q, R] = householder(A);
delta = diag(R);
sum_delta = sum(delta);
   d = log10(delta./sum_delta);
   n = length(d);
plot(1:n, d, 'b*-')
```

```
hold on
end
ylabel('log10(\delta_{\iti}/tr(\delta_{\iti}))')
xlabel('{\itd} = 3:5:23')
axis([0 25–15 0])
```

Figure 1a shows the symplectic geometry spectrums *SGS* of x without noise according to the above equations based on symplectic geometry theory. We can see that the symplectic geometry spectrums turn abruptly into a flat area from $i = 6$, i.e. $\sigma_1 > \sigma_2 > \ldots > \sigma_5 >> \sigma_{5+1} \geq \ldots \geq \sigma_d$. So, the embedding dimension of the time series x can be estimated at 6. But from the **Figure 1b**, we can see that it is difficult for the SVD method to determinate the embedding dimension from the time series x. The results indicate that the symplectic geometry method could better determinate the embedding dimension from a time series due to its preserving-measure properties.

Case 2: Embedding dimension estimation for the surface EMG signal [5].

In the practical engineering research, a lot of time series data due to their complexity are considered to be nonlinear, such as the surface EMG signal in biomedical engineering. As a kind of non-invasive measure for the contracting skeletal muscles, the surface EMG signal reflects some information about the muscle, limb movements and loading of the bones and joints. It has been widely applied to assess biomechanical and motor control deficits and other functional disorders, as well as to diagnose neuromuscular problems. However, due to noise interference, the study of surface EMG signal is still a great challenge in biomedical engineering. Many researches indicate that the surface EMG signal is complex and nonlinear. The embedding dimension estimation of the surface EMG signal is usually critical to analyze its nonlinear features. As an example, we use the above symplectic geometry method to estimate the embedding dimension of the surface EMG signal during forearm supination. The length of the surface EMG signal is 1000 points. The data sampling frequency is 1 kHz. **Figure 2a** shows the raw surface EMG signal. **Figure 2b** gives the symplectic geometry spectrums *SGS* of the data in **Figure 2a**. From **Figure 2b**, the symplectic geometry spectrums *SGS* change slowly at $d = 6$ and turn into noise floor with the increase of the index i. Then, the embedding dimension can be estimated at 6 for the surface EMG signal during forearm supination.

Case 3: SymEn analysis of vibration signals on rolling bearings [11].

In the rotating machinery systems, it is extremely important for rolling bearings to detect faults from vibration signals. The Case Western Reserve University (CWRU) Bearing Data Center provides a website database for the vibration signals

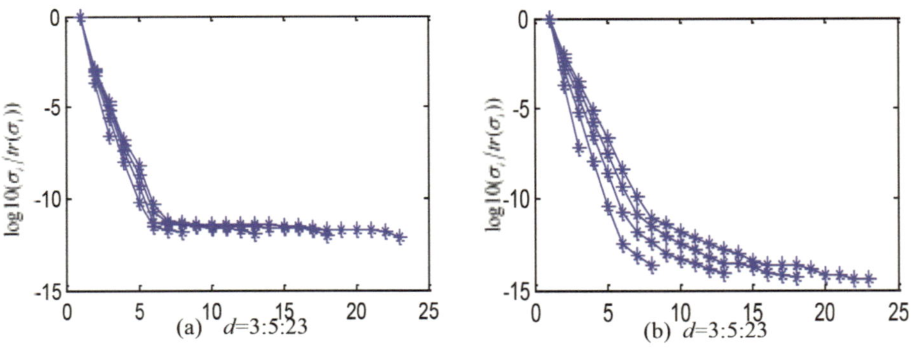

Figure 1.
The embedding dimension estimation of Lorenz chaos series with no noise based on: (a) the symplectic geometry method; (b) the SVD method.

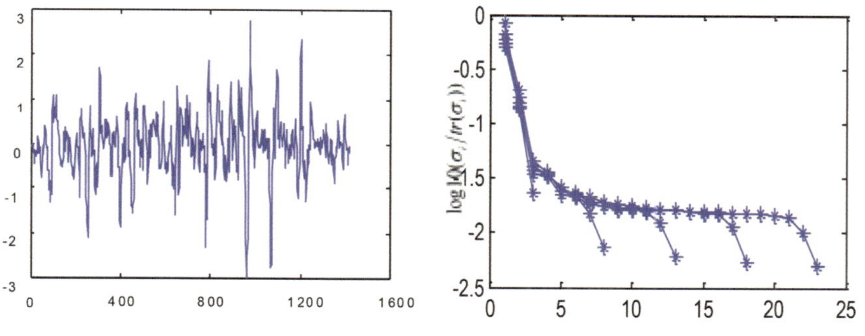

(a) (b)

Figure 2.
The embedding dimension analysis of the surface EMG signal based on the symplectic geometry spectrums: (a) Typical surface EMG signal during forearm supination; (b) The symplectic geometry spectrums of the surface EMG data in (a), where abscissa is the analysis dimension $d = 3, 8, 13, 18, 23$, *ordinate is* $SGS_i = \log(\sigma_i/tr(\sigma_i))$, *where the index* $i = 1{:}\ d$.

of bearings (http://csegroups.case.edu/bearingdatacenter /pages/welcome-case-western-reserve-university-bearing-data-center-website). From the website, the acceleration vibration data sets for 6205-2RS JEM of SKF deep-groove ball bearings are obtained to detect their fault categories. The corresponding sampling frequency is 12 kHz, the shaft speed 1730 r/min. The analyzed data sets include No.100 for normal condition(NC), No.212 for inner race fault (IRF), No.225 for rolling element fault (REF), and No.261 for outer race fault (ORF) at 12 o'clock position. The data of each set consist of the vibration signals at the housing of the drive end (DE) bearing and that of the fan end (FE) bearing, which the faults are at the drive end. The corresponding fault depth and diameter are 0.21 inches and 0.53 mm, respectively.

Symplectic geometry preserves the nature of a dynamic system under symplectic similar transformations. As an entropy measure in symplectic geometry, the SymEn value of a time series measures the lack of information in a dynamic system to reflect its properties. For the complexity of a rolling bearing, the SymEn estimate is applied to test its nonlinear characteristics from the vibration signals. **Figure 2** shows the SymEn values of the vibration signals at the drive end and their surrogate data sets based on the null hypothesis of a Gaussian linear stochastic process. Here, the length of each data is 6000 points. The embedding dimension $d = 7$.

Meanwhile, the 39 sets of surrogate data are generated by the iterated amplitude adjusted Fourier transform (IAAFT) algorithm in the 95% confidence level [26]. From **Figure 3**, we can see that there are the significant differences between these SymEn values of the vibration signals of a rolling bearing and their surrogate data sets. The results indicate that the vibration data could contain nonlinear characteristics. The original vibration signals are not from a Gaussian linear stochastic process in the 95% confidence level but from a nonlinear dynamical system. It conforms that the rolling bearing system is a complex nonlinear dynamical system.

Due to the complexity of rolling bearings, it is often thought that the high dimensional features can better identify the faults of rolling bearings [27–29]. However, the SymEn method can availably extract the low-dimensional features to identify the faults of rolling bearings from vibration signals quite precisely. **Figure 4** shows the four working states of rolling bearings, i.e., NC, ORF, REF, and IRF, based on 2-dimensional features. The abscissa is the SymEn estimates of vibration signals at the drive end. The ordinate is those estimates of vibration signals at the fan end. We can see that the four states are obviously different.

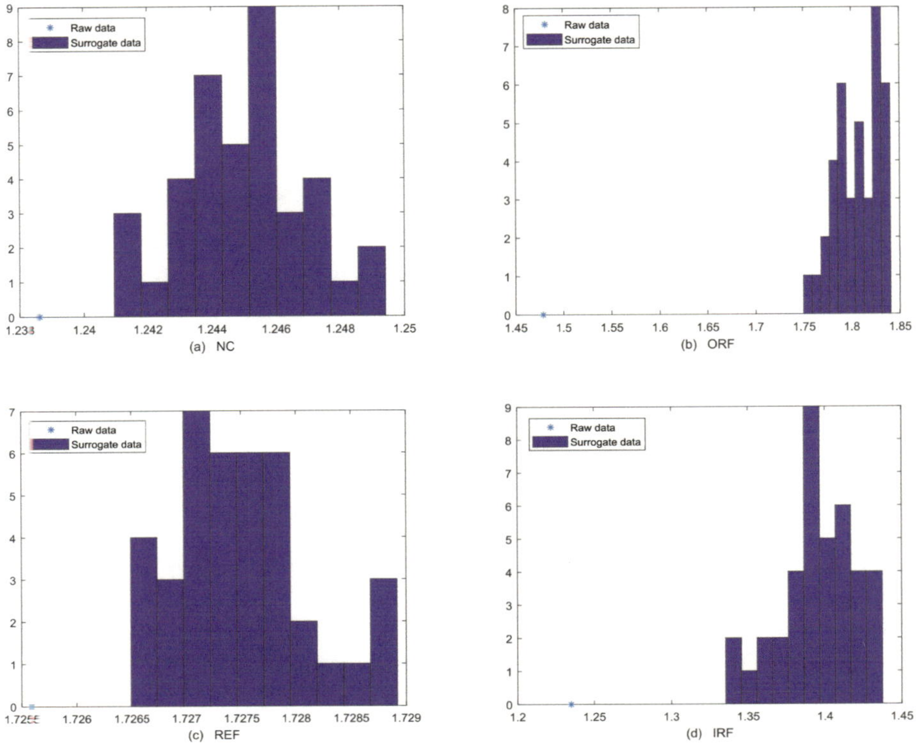

Figure 3.
The nonlinear analysis of vibration signals based on the SymEn method: (a) for the normal condition (NC); (b) for the outer race fault (ORF); (c) for the rolling element fault (REF); (d) for the inner race fault (IRF). The abscissa is the SymEn values of vibration signals and their surrogate data.

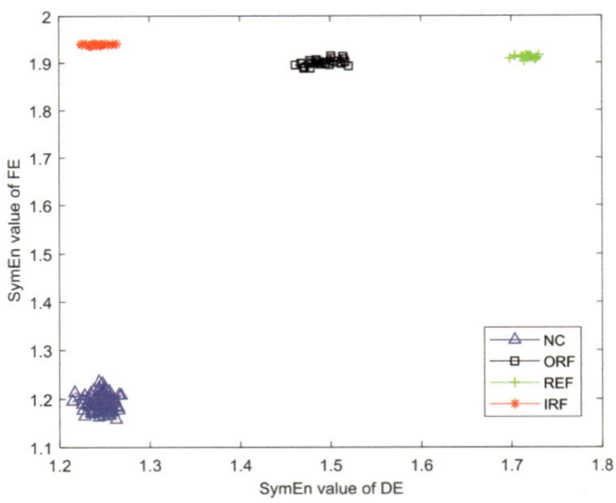

Figure 4.
The states analysis of rolling for bearings with the SymEn estimates.

There are 100% accuracies by RBF classifier for the four states of the rolling bearings. **Figure 5** plots the histogram of error values between output classes and target classes for the SymEn estimates as features of vibration signals.

Case 4: Noise reduction analysis of vibration signals based on SPCA [17, 30].

In the practical engineering measurement, the vibration data of rolling bearings have often become contaminated with noise. The noise reduction is also beneficial to analyze the measured data. The SPCA method preserves the intrinsic nonlinear nature of the raw data. The symplectic principal components can better retrieve

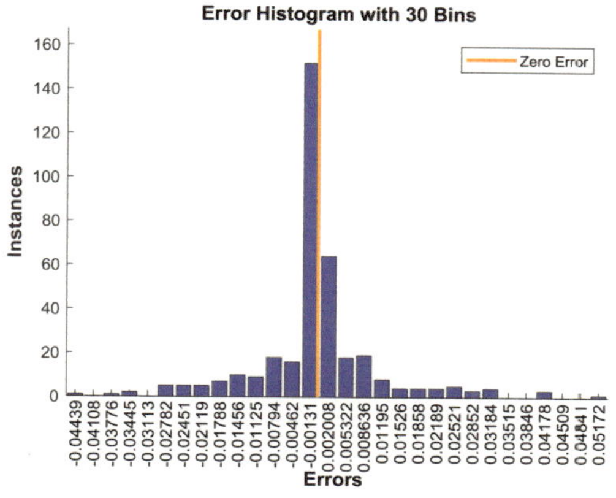

Figure 5.
The analysis of error values identification accuracies of four states.

dominant patterns from the noisy data. For the vibration signals of rolling bearings, the first symplectic principal component is used two times continuously to reduce the noise in the data.

The specific analysis procedures are as follows:

1. Build a Hamiltonian matrix from the measured data in terms of Eq. (1), Definition 2.1, 2.2 and Theorem 2.3;

2. Use the Eq. (44)–(59) to compute a symplectic Householder transform matrix Q for the symplectic QR decomposition in the SPCA method;

3. Construct the first symplectic principal component eigenvector matrix $\boldsymbol{p}_1$;

4. Calculate the first symplectic principal component coefficients $\boldsymbol{S}_1$, i.e.:

$$\mathbf{S}_1 = \mathbf{p}_1^T\mathbf{X} = \mathbf{X}^T\mathbf{p}_1;$$

5. Get the first denoised data $\boldsymbol{x}_1$ from the reestimated matrix in the following:

$$\hat{\mathbf{X}}_1 = \mathbf{p}_1\mathbf{S}_1;$$

6. Let the first denoised data $\boldsymbol{x}_1$ into the first step, and repeat the above steps, then obtain the second denoised data $\boldsymbol{x}_3$.

Figure 6 shows the effect of denoising for the vibration signals of rolling element fault (REF), No.225 data in the CWRU database [11]. For the rolling element fault at the drive end, the fault state can be seen clearly by the second reducing noise (see **Figure 6a**). For the vibration signals at the fan end without faults, the periodical characteristics in the normal state can be shown after the two reducing noise (see **Figure 6b**).

Moreover, the noise reduction method based on the symplectic geometry has been used to denoise several time series data of Lorenz chaotic system, duffing chaotic system, Chua's chaotic system with noise, as well as the sunspot number [30]. The details can be found in literatures [17, 30].

Besides, the symplectic geometry method also further integrate other approaches to better investigate the fault extraction and identification for rotating systems, such as symplectic geometry mode decomposition [19] with power

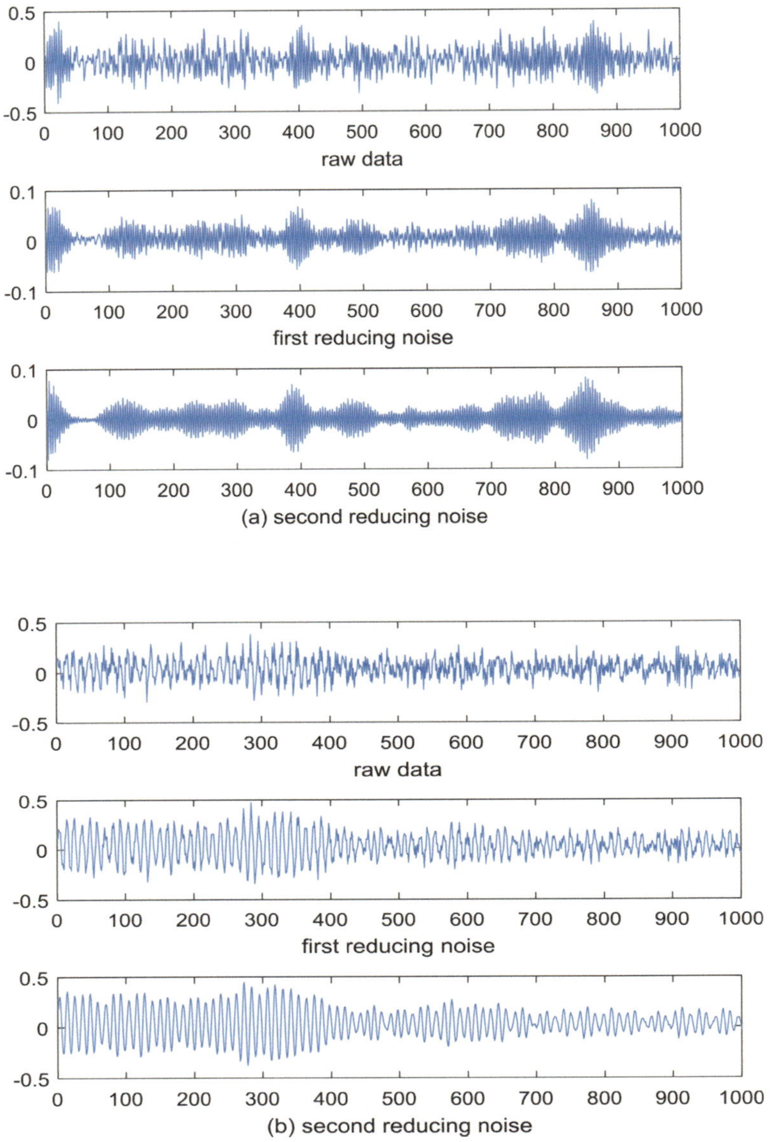

Figure 6.
The two times denoising analysis for the vibration signals of rolling element fault (REF) in No.225 data from the CWRU database. (a) The abscissa is the number of data points; (b) the ordinate is the amplitude (v) of the data.

spectral entropy [7] as well as Lagrange multiplier [20], symplectic transformation based variational Bayesian learning [21].

4. Conclusions and future research

This chapter introduces the symplectic geometry theory in the research field of the time series analysis in view of the complexity of a time series. Corresponding to Euclidean geometry, the basic concepts and basic elements of mathematics of the symplectic geometry are given, such as the symplectic space, symplectic transformation, Hamiltonian matrix, symplectic entropy (SymEn), symplectic principal component analysis (SPCA), and so on. Based on the symplectic geometry theory, the symplectic geometry spectrum analysis (SGSA), the symplectic entropy (SymEn) method and the symplectic geometry mode decomposition (SGMD) method are demonstrated to investigate the principal characteristics of a time series in the symplectic space. Meanwhile, the corresponding matlab programs are given. At last, in order to facilitate readers to learn, use and develop the symplectic geometry method, some applications of symplectic geometry on time series analysis are presented, such as the embedding dimension estimation, nonlinear testing, fault diagnosis, as well as noise reduction.

The embedding dimension estimation is often the first step in nonlinear time series analysis. Case 1 and 2 show the embedding dimension estimation of Lorenz chaotic time series and the surface EMG signal based on symplectic geometry spectrum. Moreover, the symplectic entropy method is applied to detect the nonlinearity of vibration signals on rolling bearings and identify the faults of vibration signals on rolling bearings (see Case 3). Considering the noise pollution in the practical engineering measurement, to dispose of the noise problem is very necessary for the measured time series analysis. Case 4 uses the SPCA method based on symplectic geometry to investigate the denoise of the vibration signals for rolling element fault (REF) from the CWRU database.

Symplectic geometry provides a new research idea for data analysis in practice. Although the symplectic geometry theory has been developed and applied on the nonlinear time series analysis, the related research based on symplectic geometry still needs to be further developed. Many future challenges in the research of symplectic geometry theory and various applications on a number of diverse aspects need to be developed furtherly. This chapter is only to provide a snapshot of some current trends and future challenges in the research of symplectic geometry theory on the time series analysis.

Acknowledgements

This work was supported by Shanghai "Science and technology innovation action plan" bio-medicine science and technology support project (Grant No. 19441907400).

Conflict of interest

The authors declare no conflict of interest.

Author details

Min Lei
School of Mechanical Engineering, Shanghai Jiao Tong University, Shanghai, P.R. China

*Address all correspondence to: leimin@sjtu.edu.cn

References

[1] Tao T. An Introduction to Measure Theory. Graduate Studies in Mathematics, 2011, vol.126, published by American Mathematical Society. ISBNs: 978–0–8218-6919-2 (print); 978–1–4704-1187-9 (online). DOI: http://dx.doi. org/10.1090/gsm/126.

[2] Kantz H, Schreiber T. Nonlinear time series analysis. Cambridge University Press, 1997.

[3] Wimberger S. Nonlinear Hamiltonian systems. Chaptered in Nonlinear Dynamics and Quantum Chaos, 2014; 21–102.

[4] Yao W, Zhong WX, Lim CW. Symplectic elasticity. World Scientific Press, 2009. DOI: 10.1142/6656

[5] Lei M, Wang ZZ, Feng ZJ. A method of embedding dimension estimation based on symplectic geometry. Physics Letters A. 2002; 303(2–3): 179–189.

[6] Xie HB, Dokos S, Sivakumar B, Mengersen K. Symplectic geometry spectrum regression for prediction of noisy time series. Physical Review E. 2016; 93: 052217.

[7] Zheng Z, Xin G. Fault feature extraction of hydraulic pumps based on symplectic geometry mode decomposition and power spectral entropy. Entropy. 2019; 21: 476–1-24.

[8] Jiang Y, Chen WZ, Li MY. Symplectic geometry decomposition-based features for automatic epileptic seizure detection. Computers in Biology and Medicine. 2020; 116: 103549.

[9] Takens F. Detecting Strange Attractors in Turbulence. In: Dynamical Systems and Turbulence, eds. D. A. Rand and L. –S. Young, Lecture Notes in Mathematics. Warwick, 1980, vol.898. (Springer-Verlag, New York, 1980), 366–381.

[10] Pincus SM. Approximate entropy as a measure of system complexity. Proceedings of the National Academy of Sciences USA. 1991; 88(6): 2297–2301.

[11] Lei M, Meng G, Dong GM. Fault detection for vibration signals on rolling bearings based on the symplectic entropy method. Entropy. 2017; 19: 607–1-17.

[12] Richman JS, Moorman JR. Physiological time-series analysis using approximate entropy and sample entropy. American Journal of Physiology Heart & Circulatory Physiology. 2000; 278: H2039-H2049.

[13] Skokos Ch. The Lyapunov characteristic exponents and their computation. Chaptered in Dynamics of Small Solar System Bodies and Exoplanets. 2009; 63–135.

[14] Vibe K, Vesin JM. On chaos detection methods. International Journal of Bifurcation and Chaos. 1996; 6(3): 529–543.

[15] Fraser AM. Information and entropy in strange attractors. IEEE Trans. Inf. Theory IT. 1989; 35: 245.

[16] Lei M, Meng G, Zhang WM, Wade J, Sarkar N. Symplectic entropy as a novel measure for complex systems. Entropy. 2016; 18: 412–1-18.

[17] Lei M, Meng G. Symplectic principal component analysis: A new method for time series analysis. Mathematical Problems in Engineering. 2011; 2011: 793429.

[18] Xie HB, Guo TR, Sivakumar B, Liew AW, Dokos S. Symplectic geometry spectrum analysis of nonlinear time series. Proceedings of The Royal Society A. 2014; 470: 20140409–1-17.

[19] Cheng J, Yang Y, Li X, Pan HY, Cheng JS. An early fault diagnosis method of gear based on improved symplectic geometry mode decomposition. Measurement 2020; 151: 107140.

[20] Pan HY, Yang Y, Zheng JD, Cheng JS. A noise reduction method of symplectic singular mode decomposition based on Lagrange multiplier. Mechanical Systems and Signal Processing. 2019; 133: 106283.

[21] Pan HY, Yang Y, Zheng JD, Cheng JS. Symplectic transformation based variational Bayesian learning and its applications to gear fault diagnosis. Measurement 2019; 147: 106827.

[22] Lei M, Wang ZH, Feng ZJ. The application of symplectic geometry on nonlinear dynamics analysis of the experimental data. In: 14th International Conference on Digital Signal Processing Proceeding; 1–3 July 2002; Santorini, Greece. 2002. v1–2, p.1137–1140

[23] Niu X, Qu F, Wang N. Evaluating Sprinters' Surface EMG Signals Based on EMD and Symplectic Geometry. Journal of Ocean University of Qingdao. 2005; 35(1): pp125–129. (in Chinese)

[24] Xie H, Wang Z, Huang H. Identification determinism in time series based on symplectic geometry spectra. Physics Letters A, 2005,342 (1–2):156–161.

[25] Lorenz EN. Deterministic nonperiodic flow. Journal of Atmospheric Sciences. 1963; 20: 130–141.

[26] Schreiber T, Schmitz A. Improved surrogate data for nonlinearity tests. Physical Review Letters. 1996; 77: 635–638. DOI: 10.1103/PhysRevLett.77.635.

[27] Yang J, Zhang Y, Zhu Y. Intelligent fault diagnosis of rolling element bearing based on SVMs and fractal dimension. Mechanical Systems and Signal Processing. 2007; 21:2012–2014. DOI:0.1016/j.ymssp.2006.10.005.

[28] Prieto MD, Cirrincione G, Espionsa AG, Ortega JA, Henao H. Bearing fault detection by a novel condition-monitoring scheme based on statistical-time features and neural networks. IEEE Transactions on Industrial Electronics 2013; 30: 3398–3407. DOI:10.1109/TIE.2012.2219838.

[29] Wu S, Wu C, Wu T, Wang C. Multi-scale analysis based ball bearing defect diagnostics using mahalanobis distance and support vector machine. Entropy 2013; 15: 416–433. DOI:10.3390/e15020416.

[30] Lei M, Meng G. A noise reduction method for continuous chaotic systems based on symplectic geometry. Journal of Vibration Engineering & Technologies. 2015; 3(1): 13–24.